ENERGY AND ENVIRONMENTAL POLICY MODELING

INTERNATIONAL SERIES IN
OPERATIONS RESEARCH & MANAGEMENT SCIENCE

Frederick S. Hillier, Series Editor
Stanford University

ENERGY AND ENVIRONMENTAL POLICY MODELING

edited by

John Weyant
Stanford University

Kluwer Academic Publishers
Boston/Dordrecht/London

Distributors for North, Central and South America:
Kluwer Academic Publishers
101 Philip Drive
Assinippi Park
Norwell, Massachusetts 02061 USA
Telephone (781) 871-6600
Fax (781) 871-6528
E-Mail <kluwer@wkap.com>

Distributors for all other countries:
Kluwer Academic Publishers Group
Distribution Centre
Post Office Box 322
3300 AH Dordrecht, THE NETHERLANDS
Telephone 31 78 6392 392
Fax 31 78 6546 474
E-Mail <orderdept@wkap.nl>

 Electronic Services <http://www.wkap.nl>

Library of Congress Cataloging-in-Publication Data

Energy and environmental policy modeling / edited by John Weyant.
       p.   cm. -- (International series in operations research &
  management science ; 18)
   Includes bibliographical references and index.
   ISBN 0-7923-8348-6
   1. Energy development--Environmental aspects--Mathematical models.
  2. Environmental policy--Mathematical models. I. Weyant, John P.
  (John Peter)   II. Series.
  TD195.E49E5116   1999
  333.79'14'015118--dc21   98-45651
JK   CIP

Energy and Environmental Policy Modeling

Contents

Chapter 1

INTRODUCTION AND OVERVIEW
John P. Weyant

This introduction describes the motivation for this book, introduces a few key energy-environmental modeling concepts, and provides an overview of the remaining chapters. There were two major motivations for this book. First, the application of management science techniques to energy and environment policy issues has expanded dramatically over the last twenty-five years or so, and results from these applications have become increasingly relied upon by those making decisions. Thus, an overview of some the most interesting work in this field on some of the biggest contemporary policy issues seemed like a worthy objective. Second, most of the authors of this volume participated in a 70th birthday celebration for Alan Manne of Stanford University in 1995 and thought dedicating this volume to him would be a fitting tribute to his many contributions to this field over the last several decades. Some of us have had the good fortune to have worked directly with Alan, others to have been his students, and others simply to read his papers, but we have all learned from him and been inspired by him and his work.

To some it might seem limiting to include only work directly linked to a single person in a book trying to give an overview of recent advances in a field as broad as energy-environmental policy. However, Professor Manne has made major contributions to a wide range of policy issues – oil and gas policy, electricity policy, global climate change, etc. In addition, he has contributed to the development of better algorithms for solving the problems he found, and to the development of institutions designed to enable us to learn from a range of analysts studying a particular problem. Thus, we felt even our relatively limited part of his intellectual family tree could provide a reasonably comprehensive perspective on energy – environmental modeling as currently practiced.

John Weyant (ed.), ENERGY AND ENVIRONMENTAL POLICY MODELING. Copyright © 1998. Kluwer Academic Publishers. ISBN 0-7923-8348-6. All rights reserved.

Basic Concepts

We start with a brief overview of some of the basic concepts used in energy-environmental modeling for readers who are not specialists in this field. In this section we describe seven basic concepts that are widely used in energy-industrial models. These concepts are employed in models employed in many of the models discussed subsequently. Thus, discussing these crosscutting concepts first allows us to greatly economize on our model application review. The basic concepts discussed here are: (1) resource supply and depletion modeling; (2) price and non-price determinants of energy demand; (3) interfuel competition; (4) end-use modeling; (5) integrating energy supply and demand; (6) aggregate energy economy interactions; and (7) multi-sectoral energy economy interactions.

Resource Supply and Depletion

One concept represented in models that include resource supply is the idea of an upward sloping supply curve, representing the assumption that the lower cost deposits of the resource will generally be found and produced before higher cost deposits. Conceptually this relationship is represented with a long run marginal cost curve for the resource. In a one period model, some estimate is made of how much of the resource will be used between now and the target year and a static supply curve is derived from the long run marginal cost curve.

Another feature often included in intertemporal model with resource depletion is the capability to include an intertemporal resource rent, reflecting the propensity of resource owners to delay producing the resource if their net profit (resource price less production costs) is increasing faster than their rate of return on alternative investments (usually approximated by the rate of interest) and to produce it all immediately if their net profit is increasing more slowly than the rate of return on alternative investments. Together these two conditions require that the difference between the price and cost of the resource rise at the rate of interest. Hotelling (1931) derives the implied equilibrium conditions for the case of a depletable resource with zero extraction costs; Solow and Wan (1977) add results for the case where extraction costs are non-zero; and Nordhaus (1973) uses the concept of a "backstop" technology, available in unlimited supply, but with high cost, as a natural terminal condition on the differential equations derived by Hotelling and Solow, leading to closed form solutions. Oren and Powell (1985), Krautkramer (1986), Powell and Oren (1989) analyze the implications of more complicated, but also more realistic, backstop specifications. For a good general overview of depletable resource theory and practice see Dasgupta and Heal (1979). Despite its theoretical appeal, actual increases in energy prices have generally been smaller than predicted by the pure theory, as improvements in resource discovery and recovery technologies have lead to cost reductions that have offset a significant portion of the increase in scarcity rents predicted by the theory. This is not in itself a critique of the theory, but in the way it has been applied to date.

Price and Non-Price Determinants of Energy Demand

The simplest models of energy demand project demand as a function of future prices and incomes. To summarize the relevant relationships economists often use elasticity estimates. The price elasticity of demand is the percentage reduction in energy demand in response to a one percent increase in energy prices after enough time has passed for the economy to adjust its energy using equipment and structures to the price change; the income elasticity of demand is the percentage increase in demand in response to a one percent increase in income. A simple back-of-the-envelope energy demand forecasting equation (from Manne, et al. 1979) might be:

$$\text{\% change in energy} = \begin{bmatrix} \text{income} \\ \text{elasticity} \end{bmatrix} \begin{bmatrix} \text{\% change price} \\ \text{in} \\ \text{GNP} \end{bmatrix} \cdot \begin{bmatrix} \text{price} \\ \text{elasticity} \end{bmatrix} \begin{bmatrix} \text{\% change} \\ \text{in energy} \\ \text{price} \end{bmatrix}$$

For example, suppose the long-run price elasticity for primary energy in the United States is 0.4 and the income elasticity is 1.0. Furthermore, suppose the U.S. economy is projected to grow at a rate of 2.4% per year, and it is desired to limit the growth of primary energy demand to 1% per year because of resource scarcity or environmental concerns. Then, energy prices would need to increase at (2.4 - 1.0)/.4 = 3.5% per year in inflated-adjusted terms to meet the limit on energy demand growth (for more on aggregate energy price elasticities see EMF, 1980).

There is considerable debate about the extent to which factors other than price and income need to be considered in projecting energy demand. In aggregate models, a common practice in recent years has been to add a time trend to the simple forecasting equation given above, representing a gradual shift in energy use per unit of economic activity (energy efficiency) that is independent of prices. If such a trend were an improvement in energy efficiency of .5% per year, the price increase in the above example would fall to only 2.25% per year in inflation adjusted terms. In more disaggregated models various attempts have been made to measure causal determinants of energy demand other than price and income. Factors considered include: (1) changing industry structure (discussed below); (2) changes in productivity; and (3) technological innovation (discussed below), especially the way changes in prices may drive technological innovation and changes in productivity.

Another complication to the simple aggregate demand forecasting equation shown above is the time lags in the adjustment of demand to price changes. Since energy using capital stock, e.g., industrial boilers, automobiles, is long lived, the adjustment to sharp changes in energy prices can take many years to be completed. Models that attempt to capture these dynamics generally treat the existing capital stock as completely (or nearly completely) non-adjustable while new capacity (for replacement or to satisfy growth in demand) is completely flexible. Only 5-10% of the capital stock is typically assumed to be new each year (for a slightly more flexible formulation see Peck and Weyant, 1985). Price elasticities, therefore, have to be estimated for both short-term and long-term adjustments.

Interfuel Competition

Another key aspect of the demand for a particular fuel is interfuel competition. Capturing interfuel competition is of great importance in projecting the demand for an individual fuel, and can also improve any estimate of aggregate energy demand. One simple way to capture interfuel competition is through the use of own- and cross- price elasticities of demand; the own-price elasticity is the percentage reduction in the demand for a fuel in response to a one percent increase in its own price and the cross-price elasticity of a particular fuel with respect to the price of a competitive fuel is the percentage change in the demand for that fuel in response to a one percent change in the price of the competitive fuel. Own-price effects are generally negative and cross-price effects are generally positive. As an example of the application of the cross-price elasticity concept, consider the demand for oil in the long run to be dependent on income, the price of oil and the price of competitive fuels.

Then our simple demand forecasting equation above can be rewritten as:

$$\begin{matrix} \% \text{ change} \\ \text{in oil} \\ \text{demand} \end{matrix} = \begin{bmatrix} \text{income} \\ \text{elasticity} \end{bmatrix} \begin{bmatrix} \% \text{ change price} \\ \text{in} \\ \text{GNP} \end{bmatrix} - \begin{bmatrix} \text{own} - \text{price} \\ \text{elasticity} \end{bmatrix} \begin{bmatrix} \% \text{ change} \\ \text{in} \\ \text{oil price} \end{bmatrix}$$

$$+ \begin{bmatrix} \text{cross} - \text{price} \\ \text{elasticity} \end{bmatrix} \begin{bmatrix} \% \text{ change price} \\ \text{in} \\ \text{price of comp. fuel} \end{bmatrix}$$

For example, suppose the income elasticity for oil demand is 1.0, the own-price elasticity is .6 and the cross-price elasticity with respect to the price of competitive fuels is .2. In addition, suppose the price of competitive fuels is expected to increase at 3 percent per year and energy security concerns lead policy makers to set no increase in oil consumption as an objective. Then oil prices would need to increase at a [(1.0)(2.4) + (.2)(3.0)]/.6 = 5 percent per year to keep oil consumption constant (see Sweeney, 1984, for a good overview of energy demand modeling and Bohi, 1981 for more on energy price elasticities).

End-Use Modeling

Another concept employed in most energy models is process engineering estimates of the cost and efficiency of various energy conversion processes. For example, models that consider the supply and demand for all energy fuels generally include representations of one or more electric generation technologies that use fossil fuel combustion to run a turbine generator. Such a system generally has an energy conversion efficiency of 35-40%. Such characterizations are common in process analyses employed in applied mathematical programming and they are used in some models in precisely this way. In others models, however, process engineering modules are embedded in market equilibrium networks as discussed below.

Process engineering is widely used to model primary energy conversion owing to the large scale and homogeneous design of the facilities involved (see, e.g., EMF, approach is increasingly being used to analyze energy demand technologies. By representing demand in terms of end-use energy services (e.g., industrial process heat, residential space heating, and automobile vehicle miles traveled), such models include representations of alternative technologies that can be used to satisfy those end-use demands. These technologies can have different costs, conversion efficiencies and employ different input fuels. Thus different levels of home insulation and competition between gas combustion and electric heat-pumps can be explicitly considered, leading many to refer to this approach as a "bottom up" approach to energy demand modeling, as contrasted with the "top down" approach employed in many aggregate economic analyses. The advantages of the increased technological detail included in end-use energy modeling must be balanced against the increased data requirements and lack of behavioral response estimates at the appropriate level. For example, although many estimates of the price elasticity of the demand for electricity by consumers are available, not many estimates of the end-use price elasticity for residential space heating are available. This results from the existence of publicly available data on fuel purchases, but not on end-use energy use consumption. By including technological data in consumer choice models in recent years, econometricians have starting producing credible estimates of end-use demand elasticities.

Integrating Supply and Demand

In models that cover both, energy supply and demand are balanced through the determination of a market clearing set of prices. These prices may be determined either through iteration or optimization because of the equivalence between the two approaches for the case of a single energy product (originally shown by Samuelson, 1952; see also Scarf, 1990 for a general discussion of the relationship between optimization and economic theory). In the iterative approach the equilibrium price and quantity are sought through a search procedure. In the optimization approach "net economic benefits," measured by the difference between the value consumers place on the amount of energy supplied and what it cost producers to supply it, i.e., the area between the demand curve and the supply curve for energy, are maximized. This area is maximized precisely at the market-clearing price; if more energy is produced at that point, the cost of production will exceed its value to consumers. Although it is possible to approximate the equilibrium by using step function approximations to the supply and demand functions in a linear optimization formulation, this is usually awkward computationally, so non-linear programming or complementary formulations are often employed. Ahn and Hogan (1979) show conditions under which this process will converge when the demand system is linearized for a single-period formulation. Daniel and Goldberg (1981), extend these results to a multi-period optimization formulation. Greenberg and Murphy (1985) show how to compute regulated price equilibria using mathematical programming.

Aggregate Energy-Economy Interactions

Thus far we have proceeded as if the level of economic activity has a strong influence on energy demand, but not vice versa. This is sometimes referred to as a partial equilibrium analysis. If a long time horizon must be considered to address the policy issue of interest or if energy prices are expected to increase dramatically, the impact of changes in energy prices on the level of economic activity needs to be taken into account. This requires the use a general equilibrium model in which the prices and quantities of all goods and services in the economy are allowed to vary. The simplest form of a such a model, and one that has been put to good use in energy analysis, splits all goods and services produced in the economy into just two categories - energy inputs and all other goods. Since energy and other goods are imperfect substitutes, both in production and consumption, when energy costs increase, more of the other goods are devoted to energy production <u>and</u> less energy inputs are consumed, as the substitution unfolds. Both of these effects reduce the total level of output of the economy. And less economic activity leads to less energy demand via equation 1. Since energy inputs represent only 5-10% of all inputs to the economy, the energy-economy feedback can usually be ignored for policy issues for which the appropriate evaluation period is less than 20 years unless very large changes in energy prices are envisioned (see EMF, 1977, and Hitch, 1977 for more on energy/economy interactions).

Multi-Sector Energy-Economy Interactions

Although our simple back-of-the-envelope energy demand projection formulae treated the price sensitivity of energy demand to changes in energy prices as homogeneous across all sectors of the economy, the substitution potential varies widely from sector to sector. For some applications, it may be very important to capture this heterogeneity. A first cut at the nature of these interactions can be gained through the use of standard fixed-coefficient input analysis, where the output of each sector of the economy is represented as being produced by a fixed fraction of the outputs of the other sectors and of the primary input factors. Thus, the instantaneous impact of a reduction in energy availability on the output of each sector can be estimated. Of course, if energy availability is reduced or energy prices increase, producers will substitute other goods in the production process and consumers will shift away from purchasing energy-intensive products. This realization lead to the development of interindustry models with input/output coefficients that depend on the relative prices of all intermediate goods and primary inputs (e.g., Hudson and Jorgenson, 1974). In these models the price elasticity of each fuel depends on the existing industry structure and on the price elasticity within each sector of the economy. Thus, the price elasticities of the individual fuels and the degree of substitution between energy and other goods can shift substantially over time in such models. It is also possible to formulate single sector models where energy demand depends on both fuel and non-fuel factor input prices. Such models can be used to study the dynamics of adjustment on the capital stock in that one sector to variations in all prices (see, e.g., Peck, Bosch, and Weyant, 1988).

Overview of Book

The book starts with policy studies. First there is a chapter on oil resource depletion and technological change by John Rowse. Two articles follow this on electric sector restructuring by William Hogan and Hung-Po Chao/Stephen Peck. Next there are four chapters motivated by the debate over appropriate climate change policies: one by Richard Richels, Jae Edmonds, Howard Gruenspecht, and Tom Wigley on the implications of different carbon emissions trajectories; one by Stephen Peck and Thomas Teisberg on appropriate emissions trajectories under uncertainty, one by Susan Swinehart on the potential of tree planting as a climate policy response, and one by Gunter Stephan on inter-generational discounting. The book ends with a chapter on the sequence of linear complimentarity problem technique for solving general equilibrium problems by Tom Rutherford, and one on the International Energy Network by Leo Schrattenholzer, an institution managed jointly by Schrattenholzer and Alan Manne. Thus, the work of Professor Alan Manne is heavily referenced in every chapter of this book.

References

Bohi, D.R. 1981. *Analyzing Energy Demand Behavior: A Study of Energy Elasticities*. Johns Hopkins University Press for Resources for the Future, Baltimore.

Daniel, T.E. and H.M. Goldberg. 1981. "Dynamic Equilibrium Modeling: The Canadian BALANCE Model,". *Operations Research, Vol.* **29**, 829-852.

Energy Modeling Forum (EMF). 1977. *Energy and the Economy. EMF 1 Summary Report*. Volumes 1 & 2, Stanford University, Stanford CA.

Energy Modeling Forum (EMF). 1980. *Aggregate Elasticity of Energy Demand. EMF 4 Summary Report. Volume 1. Vol. 2 (1981).* Stanford University, Stanford CA.

Energy Modeling Forum (EMF). 1987. *Industrial Energy Demand. EMF 8 Summary Report. Vol. 2 (1988).* Stanford University, Stanford CA.

Greenberg, H.J. and F.H. Murphy. 1985. "Computing Price Equilibria with Price Regulations Using Mathematical Programming." *Operations Research, Vol.* **33**, 935-954.

Hitch, C., ed. 1977. *Modeling Energy Economy Interactions: Five Approaches.* Johns Hopkins University Press for Resources for the Future, Baltimore.

Hotelling, H. 1931. The Economics of Exhaustible Resources. *Journal of Political Economics, Vol.* **39**, 137-175.

Hudson, E.A. and D.W. Jorgenson. 1974. U.S. Energy Policy and Economic Growth; 1975-2000. *Bell Journal of Economics and Management Science, Vol.* **5**, 461-514.

Krautkramer, J.A. 1986. Optimal Depletion With Resource Amenities and Backstop Technology. *Resources and Energy, Vol.* **8,** 109-132.

Krautkramer, J.A. 1986. Optimal Depletion With Resource Amenities and a Backstop Technology. *Resources and Energy, Vol.* **8,** 109-132.

Manne, A.S., R.G. Richels and J.P. Weyant. 1979. Energy Policy Modeling: A Survey. *Operations Research*, Vol. 27, 1-36.

Oren, S.S. and S.G. Powell. 1985. Optimal Supply of a Depletable Resource with a Backstop Technology: Heal's Theorem Revisited. *Operations Research, Vol.* **33**, 277-292.

Peck S.C. and J.P. Weyant. 1985. Electricity Growth in the Future. *The Energy Journal, Vol.* **6**, 23-43.

Peck, S.C., D.K. Bosch, and J.P. Weyant. 1988. Industrial Energy Demand: A Simple Structural Approach. *Resources and Energy, Vol.* **10**, 111-134.

Samuelson, P.A. 1952. Spatial Price Equilibrium and Linear Programming. *American Economic Review, Vol.* **42**, 283-303.

Scarf, H.E. 1990. Mathematical Programming and Economic Theory. *Operations Research, Vol.* **38**, 377-385.

Solow, J.L. and F.Y. Wan. 1987. Extraction Costs in the Theory of Exhaustible Resources. *Bell Journal of Economics, Vol.* **7**, 359-370. 31-37.

Sweeney, J.L. 1984. The Response of Energy Demand to Higher Prices: What Have We Learned? *American Economic Review, Vol.* **74**, 31-37.

Chapter 2

TECHNOLOGICAL ADVANCES IN RECOVERY METHODS AND EFFICIENT ALLOCATION OF A NONRENEWABLE RESOURCE[*]

John Rowse
The University of Calgary
Calgary, Alberta, Canada

Abstract: Most economists would likely agree that technological progress is central to the efficient use of a nonrenewable resource over the long term. Yet advances in petroluem recovery technologies since the mid-1970s suggest that technological progress may also be important over a time span as short as two decades. This paper examines the consequences of underestimating technological advances in recovering natural gas for the Canadian province of British Columbia using a computational model maximizing conventionally-defined social welfare. Mistakes in allocating gas over time due to understimating yield relatively small welfare losses, a result that likely generalizes to other resources.

1. Introduction

Since the OPEC-induced oil price shocks of 1973/74, much technological change has occurred in crude oil and natural gas recovery methods.[1] How is technological change regarded by the resource economics literature? Although the recent survey by Kneese and Sweeney (1993) provides little explicit discussion of technological change,[2] Dasgupta (1993, 1117) argues that:

> Technological improvements and their influence on resource substitutability are a key to understanding the economics of exhaustible resources...technological improvements will enable future generations to exploit currently *unusable* resources.

Over a long time frame, technological change in energy use and supply has been enormous. Rogner (1988, 12) discusses the past and possible future shift of

John Weyant (ed.), ENERGY AND ENVIRONMENTAL POLICY MODELING. Copyright © 1998. Kluwer Academic Publishers. ISBN 0-7923-8348-6. All rights reserved.

the primary energy source during 1850-2050 from wood to coal to oil to natural gas. Dasgupta (1993, 1112-1117) and Rosenberg (1994, Ch. 9) also provide enlightening historical perspectives.   Unfortunately, as Sweeney (1993) notes, when technological change is considered in an economic analysis, fewer analytical results can be established. Moreover, as Farzin (1995) reports, ambiguous findings may result when technological change is taken into account.

Recent studies of technological change include Rogner (1988), Lohrenz (1991), Norquist (1993), Chermak and Patrick (1995) and Farzin (1995). This work focuses on natural gas and asks: how large are the benefits of technological advances in gas recovery methods and what are the losses from*underestimating* such advances? The latter question arises because in the past many energy analysts have underestimated technological progress in petroleum recovery. The questions are approached from a conventional nonrenewable resource perspective using a complex computational model of natural gas allocation for the Canadian province of British Columbia. The model maximizes social welfare, defined as the discounted sum of consumer plus producer surplus.[3]

In brief, the findings are as follows. Under a variety of gas futures the welfare gains from technological advances in gas recovery methods are relatively small, under 6%. Alternative performance measures exhibit more substantial adjustments. Underestimating technological improvements for a decade also incurs relatively small welfare losses.   As long as reallocations are made when the correct technological advances are learned, the losses can be minimized; natural gas is not lost through defective foresight, it is simply misallocated and the mistakes can largely be overcome.

This paper is organized as follows.   Discussion is next devoted to technological advances in petroleum recovery and their prospective importance for natural gas. It is also argued that adopting a perspective on technological change cannot be avoided when studying nonrenewable resource allocation.   The computational model is described subsequently, then the results are set forth and interpreted. Lessons learned are discussed in the final section. Details of the model are provided in the Appendix.

2. Technological Advances in Petroleum and Natural Gas Recovery

For crude oil and natural gas, recovery processes are frequently characterized as conventional or nonconventional, and sometimes offshore. Distinction between categories typically is on the basis of technical criteria, and technological advances may cause nonconventional recovery methods effectively to migrate into the conventional category. Technological improvements have occurred in all types of petroleum recovery. Improvements in gas recovery have occurred as well and further advances will likely make nonconventional gas more attractive.[4] For example, gas trapped in tight strata, or tight gas, is known to exist in very large amounts in Canada and the US. Some deposits have been exploited commercially in the US for decades.

Coalbed methane, or gas from coal seams, forms a second nonconventional source. Large volumes are known to exist, but thus far little coalbed methane has proved economic. However, with US government subsidies ranging up to about $0.90 per thousand cubic feet, in the early 1990s coalbed methane provided rapidly growing supplies to US markets. These supplies declined with the subsidies but the experience demonstrated just how rapidly (over only a decade) nonconventional supplies could expand in the right price environment.[5]

Recent advances in petroleum recovery methods suggest that technological change ought to be taken into account when studying petroleum allocation over a time frame as short as two decades. But a much longer time frame is typically required to calculate user costs properly. For instance, to study *ad valorem* taxation of nonrenewable resource production, Rowse (1996) utilizes a 75-year time horizon and cites four earlier studies with horizons of half a century or more. *Thus, no sensible study of nonrenewable resource allocation can avoid allowing for technological progress. Either explicit technological advances should be assumed, or the questionable assumption adopted that no advances will occur. Moreover, to bracket plausible outcomes, the consequences of aggressive technological change should also be examined.*

3. The Computational Model and Base Case Assumptions

Analysis is performed with a model of natural gas allocation for the Canadian province of British Columbia (BC) used by Rowse (1996) to examine royalty taxation.[6] Tax issues are not considered in this work and the algebraic form of the model is set forth in the Appendix.

Specific model characteristics are: perfect foresight; a time frame of 1990-2079 (90 years), with each period one year long; a single domestic consumption location; an export location on the Canada/US border; a single BC production location distant from the consumption location and the border; domestic demand consisting of fixed demand and flexible demand; and gas supplies from existing reserves, new reserves and the backstop source

Forming part of domestic demand, fixed demand is associated with the surviving gas-using capital stock, which declines with time. Fixed demand in period t is $\psi^t Q_0$, where Q_0 is 1989 BC consumption of 225.6 petajoules (PJ) (National Energy Board (1991, 358)) and, assuming that 4% of the capital stock is retired annually, the annual survival rate ψ is 0.96. The balance of domestic demand, flexible demand is given by $QR_t = \alpha_t P_t^{-\beta}$, is the (absolute) long run price elasticity. Thus consumption associated with new capital stock is fully price responsive. Price elasticity $\beta = 1.25$ and exogenous pairs $(\overline{Q}_t, \overline{P}_t)$ anchor the QR_t. Numerical values for the $\overline{Q}_t$ rest on the assumption that at a constant price $\overline{P}_t$ of $2.50 per gigajoule (GJ) (1990), ceteris paribus, domestic demand grows at 3% annually, starting at Q_0.

Domestic demand exhibits the following behaviour: Over time, as the gas-using capital stock shrinks, flexible demand gains in importance for determining consumption in each period. Hence a given price change influences near-term demands less than later demands. This behaviour has been observed in the empirical literature on energy demands; see Sweeney (1984). Rowse (1990, 776-8) discusses this demand adjustment mechanism further. Capturing demand inertia is believed important for generating sensible results.

How should export prospects be represented? No uniquely correct way exists and a simple approach is adopted: export price e_t is set to \$2.50/GJ in 1990 Canadian dollars, for all t. Export ceilings are $\overline{E}_1 = 100$ petajoules (PJ), $\overline{E}_2 = 130$ PJ, and $\overline{E}_t = 160$ PJ, for t=3,..., 90.[7]

Principal supply assumptions are as follows. *For all existing and new gas reserves*, operating costs are \$0.60/GJ and intraprovincial transport costs are \$0.30/GJ, summing to \$0.90/GJ. Initial capital costs for existing reserves are assumed to be sunk and no additional capital costs need be incurred. Productive capacities from existing reserves are based upon data underlying projections by Canada's National Energy Board -- henceforth NEB -- of BC productive capacity in NEB (1991, 407). New reserves exhibit unit capital costs which follow the exponential function $k(W) = k_1 e^{k_2 W}$, passing through (0.0 EJ, \$0.50/GJ) and (24.0 EJ, \$6.60/GJ), where W measures cumulative reserve additions in exajoules (EJ). The first point is based upon industry data, while the second is drawn from NEB (1991, Ch. 6). k(W) is graphed as cost function 1 in Figure 1. All capital costs of new reserves are incurred prior to any production, and all gas production follows a recovery profile, which for new reserves declines at 10% annually.[8]

All existing reserves and new reserves are for conventional gas. Backstop supplies may come from several different high-cost sources: offshore west coast, tight gas, synthetic gas from coal, coalbed methane, gas from the Northwest Territories or from Alberta. All sources are aggregated to form one backstop source costing \$7.50/GJ; see NEB (1991, 140). In Figure 1, k(W) is truncated at \$6.60/GJ because -- to compete with backstop gas -- conventional gas must also bear operating and intraprovincial transport costs of \$0.90/GJ.

Commonly referred to as social welfare (SW), the objective function is the private sector present-valued sum of consumer surplus (CS) plus producer surplus (PS) arising from gas production, domestic consumption and exports.

The solution of the computational model simulates competitive market outcomes. Forming part of the solution are efficient BC prices, which consist of the sum of marginal factor supply costs, transport costs and user costs.[9]

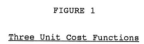

FIGURE 1

Three Unit Cost Functions

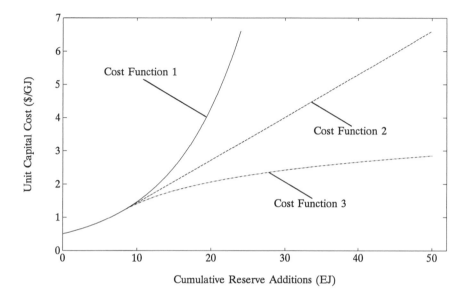

Cumulative Reserve Additions (EJ)

4. Computational Results

4.1 Base Case Outcomes

The Case 1A price and production trajectories of Figures 2 and 3 depict salient Base Case outcomes. The domestic price rises to the delivered price of backstop gas ($7.67/GJ), which exceeds the backstop cost slightly because some gas is used for pipeline fuel; see equation (A1) below. Price initially rises toward the spatial equilibrium price between the domestic and export markets, which is attained in 2001 and 2002. In each year the marginal profit of delivering gas to the export market equals that of the domestic market. Before 2001 the export market is more attractive than the domestic market and is supplied to the maximum; after 2002 it is less profitable and is thus abandoned. The Case 1A production profile -- which includes both conventional and backstop gas production -- is quite different. Production rises rapidly from 1990, tracking allowed exports, then falls sharply as exports vanish.[10]

FIGURE 2

Price Trajectories for Cases 1A, 1B and 1C

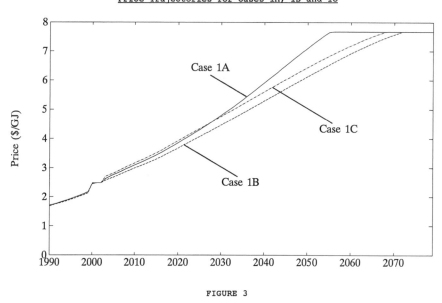

FIGURE 3

Production Trajectories for Cases 1A, 1B and 1C

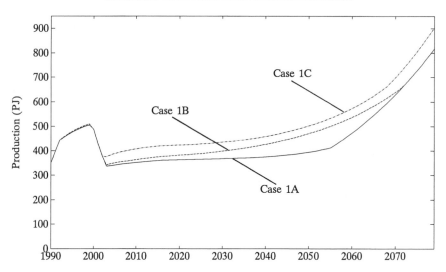

Case 1A entries in Table 1 shed light on various aspects of the solution. Domestic consumption of 37.075 EJ, domestic revenues of $21.455 billion (B) discounted to 1989, exports of 1.802 EJ, export revenue of $3.293 B, conventional production of 28.636 EJ, conventional costs of $11.562 B, backstop production of 11.148 EJ, and backstop costs of $1.620 B, may interest resource managers for measuring industry activity, job creation potential, prospective tax bases, export earnings or exchange rate implications. If backstop gas can be identified -- perhaps gas from Alberta -- then its size relative to BC supplies may be of much interest. CS of $21.121 B and PS of $11.567 B sum to SW of $32.688 B. CS uses the delivered price of backstop gas as a choke price, while PS is the sum of domestic and export revenues less conventional and backstop supply costs.[11],[12] Exports stretch through 2002, new conventional gas arrives in 2004 and backstop gas enters in 2056.

4.2 Representing Technological Progress

Technological progress slows the rate of cost increase for conventional gas and lowers nonconventional costs. It may also make gas use more efficient (altering demands over time) and gas transport less costly, but these advances are not considered. How should technological progress in gas recovery be represented?[13] There seems no unambiguously correct answer, and the following approach is adopted. After certain reserve additions unit capital cost no longer rises exponentially but follows a power function augmented by a constant:

$$h(W) = \xi_1 + \xi_2 \, W^{\xi_3}.$$

This function is appealing because it can represent a linear rise in unit cost if $\xi_2 > 0$ and $\xi_3 = 1$, a rise in unit cost and a rise in the rate of cost increase if $\xi_2 > 0$ and $\xi_3 > 1$, a rise in unit cost and a decline in the rate of cost increase if $\xi_2 > 0$ and $0 < \xi_3 < 1$, and the previous behaviour but with cost asymptotic if $\xi_2 < 0$ and $\xi_3 < 0$.

Cost function 2 of Figure 1 represents optimistic technological progress. The switchpoint to h(W) from k(W) is taken to be 8.0 EJ (one third of the 24.0 EJ specified by NEB (1991, 118)) and h(W) satisfies k(8.0 EJ) = h(8.0 EJ), k'(8.0 EJ) = h'(8.0 EJ) and h(50.0 EJ) = $6.60/GJ. The switchpoint occurs for W > 0 to allow all functions to represent the same cost information about the earliest reserve additions. Equal slopes of k and h ensure smoothness and the point (50.0 EJ, $6.60/GJ) makes 50.0 EJ of cumulative reserve additions available at the backstop cost of $7.50/GJ (less $0.90/GJ for operating and intraprovincial transport costs) instead of 24.0 EJ. Cost function 2 is almost linear, with ξ_3 computed to be 1.013.

Solving Case 1B (1A with cost function 2) yields trajectories graphed in Figures 2 and 3. Case 1B prices nearly always lie below those of 1A and the reverse is true for production levels. From Table 1, domestic consumption rises by 7.07%, domestic revenues, exports and export revenue change little, conventional production rises by 40.53%, and backstop production shrinks by 80.10%. CS rises by 5.11%, PS falls by 2.30% and SW climbs by 2.49%.

Figure 3 suggests the basic forces at work. Production patterns for Cases 1A and 1B are similar through 2020, with notable divergences for 2020 through 2065. Discounting compresses the welfare gains on production differences three decades

Table 1

Performance Measures and Comparisons for Four Cases

Case/Sub-Case	Domestic Cons'n (EJ)	Domestic Revenue (B $)	Exports (EJ)	Export Revenue (B $)	Convent'l Prod'n (EJ)	Convent'l Costs (B $)	Backstop Prod'n (EJ)	Backstop Costs (B $)	Consumer Surplus (B $)	Producer Surplus (B $)	Social Welfare (B $)	Last Year of Exports	First Year of New Conv. Prod'n	First Year of Backstop Prod'n
1A	37.075	21.455	1.802	3.293	28.636	11.562	11.148	1.620	21.121	11.567	32.688	2002	2004	2056
1B	39.696 (+7.07)	21.528 (+0.34)	1.798 (-0.22)	3.288 (-0.15)	40.242 (+40.53)	13.285 (+14.90)	2.218 (-80.10)	0.230 (-85.80)	22.200 (+5.11)	11.301 (-2.30)	33.501 (+2.49)	2002	2003	2072
1C	42.967 (+15.89)	24.194 (+12.77)	1.714 (-4.88)	3.171 (-3.70)	41.123 (+43.61)	14.567 (+25.99)	4.597 (-58.76)	0.511 (-68.46)	[25.184] (----)	12.286 (+6.22)	[37.470] (----)	2002	2003	2069
2A	35.378	22.300	0.703	1.474	24.628	11.684	12.287	1.880	18.621	10.210	28.831	1997	1990	2053
2B	37.874 (+7.05)	22.337 (+0.17)	0.898 (+27.74)	1.851 (+25.58)	36.682 (+48.94)	13.911 (+19.06)	2.988 (-75.68)	0.319 (-83.03)	19.800 (+6.33)	9.959 (-2.46)	29.758 (+3.22)	1998	1990	2071
2C	37.572 (+6.20)	22.561 (+1.17)	0.703 (0.00)	1.474 (0.00)	36.586 (+48.55)	13.737 (+17.57)	2.575 (-79.04)	0.271 (-85.59)	[21.622] (----)	10.026 (-1.80)	[31.649] (----)	1997	1990	2071
2D	64.415 (+82.08)	22.888 (+2.64)	1.263 (+79.66)	2.487 (+68.72)	22.069 (-10.39)	12.748 (+9.11)	45.129 (+267.29)	5.238 (+178.62)	[5.070] (----)	7.389 (-27.63)	[12.459] (----)	1999	1990	2034
2E	62.477 (+76.60)	22.710 (+1.84)	0.713 (+1.42)	1.489 (+1.02)	22.069 (-10.39)	11.929 (-2.10)	42.579 (+246.54)	4.843 (+157.61)	[5.106] (----)	7.426 (-27.27)	[12.533] (----)	2000	1990	2035
3A	78.125	35.179	2.984	7.429	24.640	15.952	58.356	10.969	14.149	15.687	29.836	2004	1990	2040
3B	80.490 (+3.03)	35.729 (+1.56)	3.291 (+10.29)	7.971 (+7.30)	38.300 (+55.44)	20.400 (+27.88)	47.431 (-18.72)	7.252 (-33.89)	15.524 (+9.72)	16.048 (+2.30)	31.572 (+5.82)	2005	1990	2051
3C	62.972 (-19.40)	29.570 (-15.94)	3.703 (+24.10)	8.641 (+16.31)	38.182 (+54.96)	18.903 (+18.50)	30.052 (-48.50)	4.181 (-61.88)	[16.324] (----)	15.127 (-3.57)	[31.451] (----)	2007	1990	2056
3D	201.178 (+157.51)	42.917 (+22.00)	4.587 (+53.72)	9.917 (+33.49)	22.069 (-10.43)	17.610 (+10.39)	188.461 (+222.95)	23.479 (+114.05)	[3.232] (----)	11.745 (-25.13)	[14.977] (----)	2010	1990	2022
3E	150.823 (+93.05)	34.330 (-2.41)	5.122 (+71.65)	10.576 (+42.36)	22.069 (-10.43)	16.681 (+4.57)	137.499 (+135.62)	16.222 (+47.89)	[2.817] (----)	12.004 (-23.48)	[14.821] (----)	2013	1990	2026
4A	34.919	8.030	1.194	1.929	22.033	5.685	14.920	0.085	11.011	4.190	15.201	1999	1990	2047
4B	36.321 (+4.02)	8.031 (+0.01)	1.239 (+3.77)	1.980 (+2.64)	29.528 (+34.02)	5.851 (+2.92)	8.905 (-40.32)	0.027 (-68.24)	11.112 (+0.92)	4.134 (-1.34)	15.246 (+2.96)	1999	1990	2060
4C	37.341 (+6.94)	8.378 (+4.33)	1.194 (0.00)	1.929 (0.00)	29.588 (+34.29)	6.020 (+5.89)	9.842 (-34.03)	0.032 (-62.35)	[12.067] (----)	4.255 (+1.55)	[16.322] (----)	1999	1990	2058
4D	63.876 (+82.93)	8.046 (+0.20)	1.275 (+6.78)	2.019 (+4.67)	14.937 (-32.21)	5.613 (-1.27)	51.722 (+246.66)	0.605 (+611.76)	[3.103] (----)	3.846 (-8.21)	[6.950] (----)	1999	1990	2022
4E	65.095 (+86.42)	8.324 (+3.66)	1.194 (0.00)	1.929 (0.00)	14.937 (-32.21)	5.649 (-0.63)	52.885 (+254.46)	0.675 (+694.12)	[3.126] (----)	3.929 (-6.23)	[7.055] (----)	1999	1990	2021

Notes: Each number in parentheses indicates the percentage change relative to Subcase A of the number above it. Certain measures of consumer surplus and of social welfare are not comparable between subcases; such measures are enclosed in square brackets and the corresponding percentage changes are listed as ----. Further, gas consumption, production and exports are measured in exajoules (EJ) and revenues, costs and surplus are measured in billions of dollars (B $) discounted to 1989.

and more beyond 1989. Table 1 also reveals that, although exports extend through 2002 (as before), new conventional gas arrives one year earlier and backstop gas arrives in 2072, sixteen years later.

4.3 Underestimating the Benefits of Technological Progress

Case 1B exhibits enhanced conventional supplies. Suppose, however, that information is imperfect and allocation decisions for the first ten years are based on cost function 1 of Figure 1. Then it is learned that cost function 2 applies and allocations are revised.[14] These circumstances are represented by fixing allocations for the first ten years at Case 1A levels and modifying domestic demands for 2000 and beyond using 1999 consumption for Q_0 in the revised formula $\varphi^{t-10} Q_0$ for fixed demands for $t \geq 11$. New pairs $(\overline{Q}_t, \overline{P}_t)$ anchor the new QR_t, with all $\overline{P}_t$ unchanged at \$2.50/GJ, and the new $\overline{Q}_t$ rest on demand growth of 3% annually, starting at the new Q_0.

Case 1C outcomes result, with the new trajectories graphed in Figures 2 and 3. The trajectories of 1A and 1C are identical for the first decade. Table 1 lists performance measures. *Because Case 1C demand functions from 2000 through 2079 differ from those of Cases 1A and 1B, welfare comparisons between 1C and 1A and between 1C and 1B are not possible.[15]* Some comparisons are interesting, however. Domestic consumption and revenue are larger than in 1B and exports and export revenue are smaller. Conventional production rises more than in 1B because less declining production is allocated to the post-horizon period, and backstop gas arrives earlier. PS climbs marginally.

Welfare comparisons between Case 1C and the other cases are not possible, but a different comparison is possible. Suppose that all Case 1C demand functions are known in 1989 and cost function 2 is recognized as the true cost function, but that gas use can be optimized for all years, not just for 2000-2079. This case allows measuring the welfare costs of locking into ten-year allocations based on the wrong unit capital cost function. Case 1D outcomes result, and its trajectories virtually coincide with those of 1C in Figures 2 and 3. To graph distinguishable trajectories, the axes of Figures 4 and 5 focus only on outcomes for 1990-2010. Case 1D exhibits slightly higher prices (and lower production levels) than 1C during the first decade to divert gas to the post-1999 period when demands are greater than in Cases 1A and 1B.

Table 2 lists performance measures for Cases 1D and 1C and percentage comparisons. To three decimal digits the SW loss of 1C from 1D is nil and other percentage changes are very small. Hence the loss from ten years of wrong allocations is so small as not to be measurable.

Summarizing the results, enhanced supplies of cost function 2 raise SW by \$0.813 B, but only by 2.49%. Domestic consumption increases but, more prominently, increased conventional supplies displace much backstop gas in the distant future.

Energy and Environmental Policy Modeling

FIGURE 4

Price Trajectories for Cases 1C and 1D

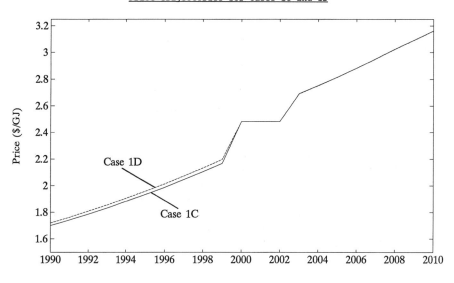

FIGURE 5

Production Trajectories for Cases 1C and 1D

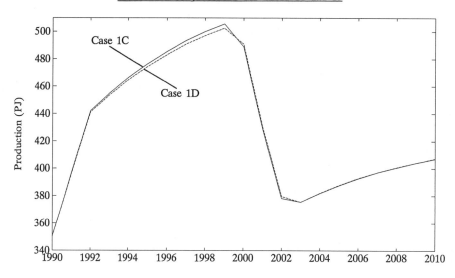

Table 2

Performance Measures and Comparisons for Selected Subcases

Sub-Case	Domestic Cons'n (EJ)	Domestic Revenue (B $)	Exports (EJ)	Export Revenue (B $)	Convent'l Prod'n (EJ)	Convent'l Costs (B $)	Backstop Prod'n (EJ)	Backstop Costs (B $)	Consumer Surplus (B $)	Producer Surplus (B $)	Social Welfare (B $)	Last Year of Exports	First Year of New Conv. Prod'n	First Year of Backstop Prod'n
1D	42.955	24.220	1.719	3.177	41.121	14.552	4.592	0.511	25.136	12.334	37.470	2002	2003	2069
1C	42.967	24.194	1.714	3.171	41.123	14.567	4.597	0.511	25.184	12.286	37.470	2002	2003	2069
	(+0.03)	(-0.11)	(-0.29)	(-0.19)	(+0.00)	(+0.10)	(+0.11)	(0.00)	(+0.19)	(-0.39)	(0.00)			
2F	37.490	22.560	0.892	1.841	36.625	14.054	2.646	0.280	21.586	10.067	31.653	1998	1990	2071
2C	37.572	22.561	0.703	1.474	36.586	13.737	2.575	0.271	21.622	10.026	31.649	1997	1990	2071
	(+0.22)	(+0.00)	(-21.19)	(-19.93)	(-0.11)	(-2.26)	(-2.68)	(-3.21)	(+1.67)	(-0.41)	(-0.01)			
2G	62.416	22.606	1.258	2.477	22.069	12.720	43.079	4.987	5.207	7.377	12.584	2000	1990	2035
2E	62.477	22.710	0.713	1.489	22.069	11.929	42.579	4.843	5.106	7.426	12.533	2000	1990	2035
	(+0.10)	(+0.46)	(-43.32)	(-39.89)	(0.00)	(-6.22)	(-1.16)	(-2.89)	(-1.94)	(+0.66)	(-0.41)			
3F	63.039	29.503	3.657	8.569	38.183	18.990	30.073	4.185	16.559	14.897	31.456	2007	1990	2056
3C	62.972	29.570	3.703	8.641	38.182	18.903	30.052	4.181	16.324	15.127	31.451	2007	1990	2056
	(-0.11)	(-0.23)	(+1.26)	(+0.84)	(-0.00)	(-0.46)	(-0.07)	(-0.10)	(-1.42)	(+1.54)	(-0.02)			
3G	151.016	34.193	5.033	10.472	22.069	16.919	137.605	16.270	3.369	11.475	14.844	2013	1990	2026
3E	150.823	34.330	5.122	10.576	22.069	16.681	137.499	16.222	2.817	12.004	14.821	2013	1990	2026
	(-0.13)	(-0.40)	(+1.77)	(+0.99)	(0.00)	(-1.41)	(-0.08)	(-0.30)	(-16.38)	(+4.61)	(-0.15)			
4F	37.370	8.375	1.133	1.860	29.586	5.955	9.811	0.032	12.074	4.248	16.322	1999	1990	2058
4C	37.341	8.378	1.194	1.929	29.588	6.020	9.842	0.032	12.067	4.255	16.322	1999	1990	2058
	(-0.08)	(+0.04)	(+5.38)	(+3.71)	(+0.01)	(+1.09)	(+0.32)	(0.00)	(-0.06)	(-0.16)	(0.00)			
4G	65.091	8.315	1.250	1.992	14.937	5.709	52.939	0.680	3.138	3.917	7.055	1999	1990	2021
4E	65.095	8.324	1.194	1.929	14.937	5.649	52.885	0.675	3.126	3.929	7.055	1999	1990	2021
	(+0.01)	(+0.11)	(-4.48)	(-3.16)	(0.00)	(-1.05)	(-0.10)	(-0.74)	(-0.38)	(+0.31)	(0.00)			

Notes: Each number in parentheses indicates the percentage change relative to Subcase D, F or G of the number above it. Further, percentage changes to social welfare are all negative, even though a nil percentage change is recorded for three subcases; recording social welfare only to three decimal digits does not allow a negative percentage decline to be reported for all cases. Finally, gas consumption, production and exports are measured in exajoules (EJ) and revenues, costs and surplus are measured in billions of dollars (B $) discounted to 1989.

If technological change is imperfectly foreseen, then after a decade the gas-using capital stock differs from its perfect-foresight level and altered demand functions lead to different optimal allocations. When comparisons are made between welfare-comparable subcases, all percentage differences are modest and the SW loss is minuscule. Thus the model is adept at minimizing welfare loss when new information is learned. Some consumers (and producers) gain from the reallocations, while other consumers (and producers) lose, but with the conventional welfare function these gains and losses largely cancel. SW distinguishes poorly among gainers and losers.

4.4 Outcomes with Reduced Existing Reserves

Deregulation of Canadian gas markets, which began in the late 1980s, left BC with a large inventory of existing reserves drilled to meet regulatory requirements for gas exports. Hence, in Case 1A new conventional gas arrives only in 2004. Thus the benefits of technological change in reducing recovery costs are shifted into the future, where discounting shrinks them. To reduce the shrinkage due to the one-time event of deregulation, Case 2A is formulated from 1A: 4.1 EJ of existing reserves, one half of the existing reserves, are arbitrarily removed. Accordingly, production capacities of existing reserves BO_1, BO_2, .., BO_{90} of equation (A2) below are reduced by 50%.[16]

The Case 2A trajectories are graphed in Figures 6 and 7. Cases 2A and 1A are comparable and the Case 2A price trajectory rises above that of 1A; see Figure 2. From Table 1, CS and PS decline and SW shrinks to $28.831 B from $32.688 B, or 11.80%. New conventional gas enters in 1990, and exports terminate five years earlier. Backstop gas arrives only 3 years earlier.

Case 2B outcomes result when cost function 2 is used. From Table 1, exports and export revenue rise noticeably and conventional production climbs by 48.94%, more than in 1B. SW climbs by $0.927 B or 3.22%. On balance the percentage differences between Cases 1 and 2 are small; thus the comparative outcomes are similar whether 4.1 EJ of existing reserves are removed or not.

What happens when the supply circumstances of cost function 2 are learned only after a decade and allocations during 1990-1999 are fixed at those of Case 2A? Case 2C provides answers, and embodies demand functions for 2000 and beyond which use 1999 consumption as the new Q_0. Figures 6 and 7 show that the Case 2B and 2C trajectories are very similar, indeed the different price trajectories cannot easily be distinguished. From Table 1 most measures are similar, save for exports and export revenues.

CS and SW are not comparable between Case 2C and the other subcases, so Case 2F is solved: the Case 2C model is used, but allocations for the first decade are not fixed at those of 2A. Table 2 shows that forcing the Case 2A model to choose decade-long wrong allocations is not very costly: SW declines by $0.004 B or 0.01% from 2F. These results confirm that the conclusions of Case 1 are little changed when 4.1 EJ of existing reserves are extinguished.

FIGURE 6

Price Trajectories for Cases 2A, 2B and 2C

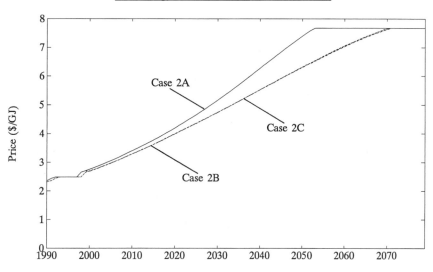

FIGURE 7

Production Trajectories for Cases 2A, 2B and 2C

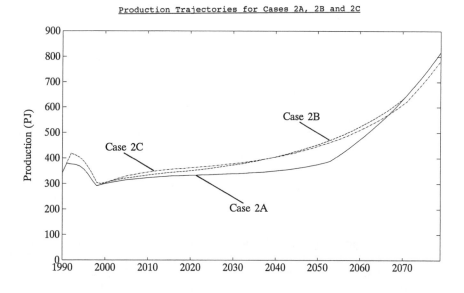

Although optimistic, cost function 2 of Figure 1 may be conservative. Suppose that aggressive technological change yields cost function 3, which rises more slowly than cost function 2 and passes through (50.0 EJ, $2.85/GJ). $2.85/GJ corresponds to a halved backstop cost of $3.75/GJ, reduced by operating and intraprovincial transport costs of $0.90/GJ.[17] For cost function 3 backstop gas will play a larger role because it can undercut the conventional supplies of cost functions 1 and 2 for prices above $2.85/GJ.

Case 2D is Case 2A with cost function 3. Its trajectories in Figures 8 and 9 contrast sharply with those of 2A. Price climbs to the new delivered price of backstop gas ($3.84/GJ versus $7.67/GJ before), backstop gas enters sooner and production increases much more rapidly after about 2010. From Table 1, domestic consumption increases by 82.08% but domestic revenues rise only by 2.64%, exports climb by 79.66% and export revenue by 68.72%, conventional production slides by 10.39% but conventional costs increase by 9.11% (conventional production occurs sooner), backstop production leaps by 267.29%, backstop costs rise only by 178.62%, PS shrinks by 27.63% and backstop gas enters in 2034. CS and SW cannot be compared with the other subcases because the delivered price of the less expensive backstop gas forms the new choke price for measuring CS. The much lower social welfare of $12.459 B is due largely to the smaller choke price.

What happens if cost function 3 is recognized as the correct cost function after a decade of allocations based on cost function 1? Case 2E provides answers, with the Case 2A 1999 domestic consumption set equal to Q_0 in the demand functions for 2000 and beyond. Figures 8 and 9 show how closely Case 2E tracks 2D, and in general the Table 1 entries support this tracking.

Case 2G provides a welfare comparison for Case 2E. The model of Case 2G is that of 2E, but Case 2A allocations are not imposed for the first decade. From Table 2 most percentage differences are modest, save for exports and export revenue, and the welfare loss climbs to 0.41%, still small.

4.5 Outcomes with Faster Demand Growth and Enhanced Export Prospects

Discounted benefits of technological advances shrink when gas supplies relying on improved recovery methods are deferred. Hence increasing early consumption may raise the benefits. To increase early consumption, the Case 3A model modifies Case 2A: 4.1 EJ of existing reserves are still removed, but domestic demands grow at 5% annually instead of 3% (at the constant price $\overline{P}_t$ of $2.50/GJ for all t); all export prices rise to $3.50/GJ from $2.50/GJ and all export ceilings rise by 50%; survival rate ☼ shrinks to 0.90 from 0.96; and price elasticity ℯ rises to 1.5 from 1.25. The latter two changes make price-responsive demands more important for determining consumption levels.

The Case 3A price trajectory of Figure 10 is similar to the Case 2A trajectory of Figure 6 except price starts higher, the spatial equilibrium price is higher and attained later, and the delivered price of backstop gas is attained sooner. The Case 3A production trajectory of Figure 11 mirrors these changes and shows their size: Case 3A production in 2079 is nearly 3500 PJ versus slightly more than 800 PJ for

FIGURE 8

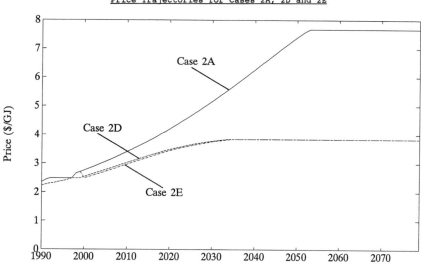

Price Trajectories for Cases 2A, 2D and 2E

FIGURE 9

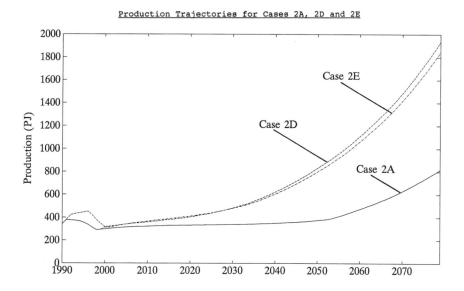

Production Trajectories for Cases 2A, 2D and 2E

FIGURE 10

Price Trajectories for Cases 3A, 3B and 3C

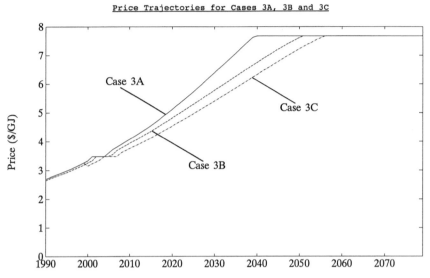

FIGURE 11

Production Trajectories for Cases 3A, 3B and 3C

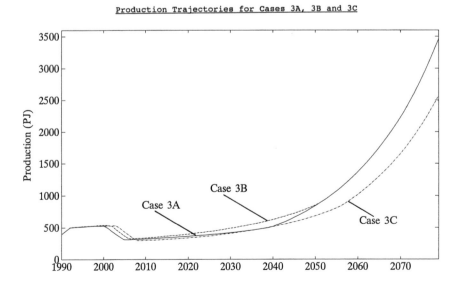

2A. Table 1 entries for 3A characterize the very different solutions. SW is not comparable between Cases 2A and 3A.

If cost function 2 applies instead of cost function 1, then the Case 3B trajectories of Figures 10 and 11 result, and SW climbs to $31.572 B or 5.82%. Larger than before, this percentage increase is still relatively small.

What happens if allocations follow those of 3A for a decade, then it is learned that cost function 2 applies? Case 3C yields answers. To provide a welfare comparison Case 3F is solved: the Case 3C model is used, but first-decade allocations are not fixed at those of 3A. From Table 2 the SW loss of Case 3C from 3F is very small at 0.02%, and other changes are small as well.

Suppose that cost function 3 applies instead of cost function 1 and that case 3A demands still apply. Case 3D outcomes result, and these diverge widely from 3A. Case 3D production in 2079 approaches 10,000 PJ, versus nearly 3500 PJ for 3A. Table 1 entries for Case 3D confirm the large changes.

What mistakes are made if allocations for the first ten years are those of 3A and at the end of 1999 it is learned that cost function 3 applies? Case 3E provides answers. From Table 1, some differences between 3D and 3E are substantial. To find a solution that is welfare-comparable, the model of 3D is solved but allocations for the first ten years are not fixed at 3A levels. Outcomes for Case 3G result, and Table 2 indicates the relatively small differences between 3G and 3E, with the welfare loss standing at only 0.15%.

4.6 Outcomes with a Higher Discount Rate

For Case 4 the Case 2 models are solved with a 10% discount rate instead of 5%. Table 1 reveals outcomes broadly similar to those of Case 2 and Table 2 reveals that SW losses from decisions using wrong information are nil when SW is recorded only to 3 decimal digits. Overall, the higher discount rate weights future outcomes less and squeezes different SW measures together, reducing the present-valued benefits from technological advances.

5. On the Generality of the Results

Cases 1-4 span a wide spectrum of futures. Extinguishing one half of existing reserves, using much more buoyant domestic demands and export prospects and doubling the discount rate lead to similar outcomes when comparing subcase A (a base case) and subcase B (optimistic technological change): percentage welfare gains are small, less than 6%. Subcase D, involving aggressive technological change, is not welfare-comparable with subcase A under any circumstances. However, making first-decade allocations using wrong information yields slight percentage losses, less than 1.0%.

Three model elements appear important for shaping allocations: near-term demands which exhibit inertia, causing near term consumption not to change much even with large price changes; existing reserves which do not benefit from technological change because their capital costs are sunk and they utilize "old" technology; and technological progress which advantages the most those reserves which are farthest along the cumulative-reserve-additions cost curve. Together these elements defer the benefits of technological advances and discounting reduces their size. Further, making wrong allocations for a decade incurs relatively small losses: only a fraction of economically recoverable reserves are consumed in a decade, and only divergences from optimal allocations are important for reducing welfare. Welfare losses are largely compensated by reallocations later.

These elements depart from reality somewhat but they embody plausible approximations. Their combined influence also suggests that the findings of relatively small welfare benefits of technological improvements and small percentage welfare losses from underestimating technological change are likely more general than the specific functional forms and parameterizations of the model indicate. Moreover, since the elements apply approximately to other nonrenewable resources, the findings may apply to other resources as well.

What circumstances might overturn the findings? Using a low discount rate could certainly raise welfare. Further, technological improvements which rapidly drive down costs and bring benefits into the near future could do so as well. Yet the inertia associated with gas use and supply systems suggest that this outcome is unlikely.[18] Moreover, in order for the losses from wrong allocations to be large, the near-term benefits would have be large and decision makers ignorant for a long time, circumstances which appear unlikely.

6. Concluding Remarks

Most resource economists would likely agree that technological change is central to the efficient allocation of a nonrenewable resource over the long term. Moreover, the petroleum industry has witnessed substantial improvements in recovery methods since the mid-1970s, suggesting that technological change may not just be a long-term phenomenon. Further, studies of nonrenewable resource allocation require long time horizons -- five decades or more -- for proper computation of user costs. Hence, technological change ought to form a part of the study of any nonrenewable resource.

Utilizing a conventional nonrenewable resource model featuring numerous complexities and focusing on supply-side technological change, this paper addresses two questions: What are the welfare gains from technological advances in gas recovery? What are the welfare losses of underestimating technological advances for a decade, then learning the correct information about incremental costs? The answer to the first question is that percentage discounted benefits are small, under 6%, although not all gains can be quantified. Other performance measures exhibit much larger percentage changes. For the second question, the percentage losses are smaller still, under 1%. When the correct information is learned, a revised allocation is found which nearly offsets losses from wrong allocations. Issues of income redistribution arise when allocations are revised, but societal attitudes toward income redistribution are not well represented by the welfare function.

Elements that appear important for shaping allocations are: inertia in gas use, that makes near-term demands relatively price non-responsive; technological progress that does not benefit existing reserves; and reserves benefitting the most from technological improvements that lie farthest along the cumulative-reserve-additions cost curve. These elements defer the greatest benefits from technological advances by several decades and discounting reduces their size. Because these elements apply regardless of the functional forms and parameterizations chosen, the findings would not likely change for many other plausible scenarios. Alternative gas futures might raise the benefits of technological improvements, but -- aside from using a smaller discount rate -- these futures would likely involve early and substantial (and thus implausible) diffusion of incremental supplies with much reduced costs. Other futures could increase the losses from decisions based on wrong information, but these futures would likely involve longer-lasting ignorance on the part of decision makers, reducing their plausibility.

Overall, the findings for natural gas are believed to be more general than the specific results suggest. Further, a plausible conjecture, given that many nonrenewable resources share these elements with natural gas, or at least approximations to these elements, is that the findings likely apply to other nonrenewable resources as well.

Any work on technological progress involves very long time frames and many assumptions. Thus qualifications are essential. The analysis of this work ignores many technological improvements extending across the chain of gas supply, pipelining and use, and hence possibly larger benefits. Moreover, the approach utilizes a conventional partial-equilibrium nonrenewable resource framework; a general equilibrium framework might yield different findings.

APPENDIX

Details of the Computational Model

Analysis is performed using a partial equilibrium model of natural gas allocation formulated by Rowse (1996). Only skeletal details are provided; Rowse (1996) provides further discussion. The time frame is 1990-2079 (90 years) and each period is one year long. All variables are non-negative.

Balance constraints equate demands to supplies:

$$(1+\epsilon)\, Q_t + (1+\nu)\, E_t = U_t + B_t + \sum_{k=1}^{t} a_{t-k+1} X_k, \quad t = 1, ..., T \qquad (A1)$$

where $T = 90$. Gas demands consist of domestic consumption Q_t and exports E_t, augmented (by parameters $\epsilon = 0.023$ and $\nu = 0.030$, respectively) to account for gas used as intraprovincial pipeline fuel. Supplies can come from U_t, the production from existing reserves, B_t, production from the backstop source, and the final term in (A1), which represents production from new reserves.

Production from existing reserves comes from developed reservoirs and satisfies constraints (A2) below. Backstop supplies come from a high-cost source with Base Case cost of $7.50/GJ in 1990 dollars.

Production from new reserves follows a thirty-year recovery profile specified by $a_1, a_2, \ldots a_T$.[19] This profile assumes 10% annual decline prior to abandonment and embodies geologic, engineering and economic factors that stretch production from reserves over many years. Parameter $a_1 = 1$, and resource commitment variable X_t (see Chao (1981)) represents the capacity of new reservoirs which first produce during year t. the sum on the right hand side of (A1) represents supplies from capacities of years t and earlier.

Gas supplies from existing reserves are required to satisfy:

$$U_1 + S_1 = BO_1 \tag{A1}$$

$$U_2 + S_2 = BO_2 + \sigma_1 S_1$$
$$U_2 + S_2 = BO_2 + \sigma_1 S_1$$

$$U_3 + S_3 = BO_3 + \sigma_2 S_1 + \sigma_1 S_2 \tag{A2}$$
$$U_T + S_T = BO_T + \sigma_{T-1} S_1 + \sigma_{T-2} S_2 + \ldots + \sigma_1 S_{t-1}$$

Nonproduced gas S_t can augment future production, but only in a way consistent with the characteristics of producing reservoirs and processing capacity. Parameters $\sigma_1, \sigma_2, \ldots, \sigma_T$ specify the production profile for nonproduced or shut-in gas, which assumes 10% annual recovery of shut-in volumes. Parameters $BO_1, \ldots BO_T$ specify productive capacities from existing reserves.

Supplies from new reserves cannot exceed exogenous bounds BN_t:

$$\sum_{k=1}^{t} a_{t-k+1} X_k \le BN_t, \qquad = 1, \ldots, T \tag{A3}$$

These bounds allow inclusion of expert opinions on how rapidly new capacity can be installed. Straitjacketing the supply response is undesirable, however, and bounds BN_t are taken so large that they never bind.

Gas reserves Y_t are related to resource commitments as follows:

$$Y_t = [\sum_{k=1}^{T} a_k] X_t, \qquad t = 1, \ldots, T \tag{A4}$$

New gas reserves cannot exceed the exogenous stock R:

$$\sum_{t=1}^{T} Y_t \le R, \tag{A5}$$

Included for completeness, this constraint is redundant because costs will rise enough to choke off gas from new reserves prior to stock exhaustion.

Exports E_t are bounded as follows:

$$E_t \le \overline{E}_t \qquad t = 1, \ldots, T \tag{A6}$$

BC demand consists of fixed demand $\psi^t Q_0$ and price-flexible demand QR_t:

$$Q_t = \psi^t Q_0 + QR_t, \qquad t = 1, ..., T \qquad (A7)$$

Survival rate ψ $(0 < \psi < 1)$ and consumption level Q_0 are exogenous. $QR_t = \alpha_t$ $P_t^{-\beta}$, where β is the (absolute) long run price elasticity.

Objective function Ω is the present value of consumer plus producer surplus arising from gas production, domestic consumption and exports:

$$\Omega = \Sigma_{t=1}^T \delta_t \Lambda_t(Q_t) + \Sigma_{t=1}^T \delta_t e_t E_t + \Sigma_{t=1}^T s_t S_t - \Sigma_{t=1}^T \delta_t u_t U_t$$

$$(A8)$$

$$- \Sigma_{t=1}^T \delta_t x_t X_t - \Sigma_{t=1}^T \delta_t P_B B_t - \overset{\wedge}{\phi}(Y_1, Y_2, ..., Y_T)$$

where δ_t is the discount factor $1/(1+r)^t$ and r is the real discount rate (5%).

Discounting is to 1989 and all costs, revenues and benefits are measured in 1990 Canadian dollars. Define the terms in (A8) as follows: $\Omega \equiv A + B + C - D - E - F - G$.

Term A is present-valued gross domestic consumer surplus, with the delivered price of backstop gas forming a choke price to measure it.[20] B is present-valued export revenues, with e_t the exogenous export price. C is the salvage value of gas from conventional reserves not produced by 2079.[21]

Remaining entries are real resource costs. D represents supply costs of existing reserves, with u_t the unit cost of extraction, processing and transport. Capital costs are sunk and thus excluded. E measures supply costs of new reserves, with x_t representing the same costs as u_t but also the recovery profile and any salvage value. F consists of backstop costs and G measures capital costs. Unit capital cost $k(W) = k_1 e^{k_2 W}$, where W represents cumulative reserve additions. Letting $\varphi(W)$ represent the integral of k(W) from 0 to W, *in a static framework* social supply costs would be:

$$\varphi(W) = k_1 e^{k_2 W} / k_2 + k_3 \qquad (A9)$$

where k_3 is chosen to satisfy $\varphi(0) = 0$. However, *in a dynamic framework*, postponing capital outlays shrinks discounted costs. Letting $W_t = \Sigma_{k=1}^t Y_k$ and using (A9), discounted social supply costs (which form term G) are:

$$\overset{\wedge}{\varphi}(Y_1, Y_2, ..., Y_T) = \delta_1 \varphi(Y_1) + \delta_2 [\varphi(Y_1 + Y_2) - \varphi(Y_1)] + ...$$

$$+ \delta_T [\varphi(\Sigma_{t=1}^T Y_t) - \varphi(\Sigma_{t=1}^{T-1} Y_t)] \qquad (A10)$$

When the unit capital cost function is h(W), $\varphi(W)$ is modified appropriately.

References

Chao, H., 1981, Exhaustible resource models: The value of information, Operations Research 29, 903-23.

Chermak, J. and R. Patrick, 1995, Technological advancement and the recovery of natural gas: The value of information, Energy Journal 16, 113-135.

Dasgupta, P., 1993, Natural resources in an age of substitutability, Chapter 25 of Kneese and Sweeney (1993).

Farzin, Y., 1995, Technological change and the dynamics of resource scarcity measures, Journal of Environmental Economics and Management 29, 105-120.

Kalisch, R. and C. McGill, 1990, Changes in natural gas recovery technology and their implications, American Gas Association Gas Energy Review 18, 2-8.

Kneese, A. and J. Sweeney (eds.), 1993, Handbook of Natural Resource and Energy Economics, Volume III, Amsterdam, North-Holland.

Lohrenz, J., 1991, Horizontal oil and gas wells: the engineering and economic nexus, Energy Journal 12, 35-53.

National Energy Board, 1991, Canadian energy: Supply and demand 1990-2010 (Supply and Services Canada, Ottawa, Ont.).

National Energy Board, 1993, Natural gas market assessment: Canadian natural gas market mechanisms - recent experiences and developments (Supply and Services Canada, Calgary, Alta.).

Norquist, S., 1993, Optimal r&d for nonrenewable resources: The case of cost reducing technology, Natural Resource Modeling 7, 219-243.

Rogner, H.-H., 1988, Technology and the prospects for natural gas: Results of current gas studies, Energy Policy 16, 9-26.

Rosenberg, N., 1994, Exploring the black box: Technology, economics, and history, (Cambridge University Press, New York).

Rowse, J., 1986, Measuring the user costs of exhaustible resource consumption, Resources and Energy 8, 365-392.

Rowse, J., 1990, Discount rate choice and efficiency in exhaustible resource allocation, Canadian Journal of Economics 23, 772-790.

Rowse, J., 1996, On *ad valorem* taxation of nonrenewable resource production, Unpublished working paper, Department of Economics, University of Calgary.

Sweeney, J., 1984, The response of energy demand to higher prices: What have we learned? American Economic Review 74, 31-37.

Sweeney, J., 1993, Economic theory of depletable resources: An introduction, Chapter 17 of Kneese and Sweeney (1993).

Endnotes

* Financial support for this work has been provided by the Social Sciences and Humanities Research Council of Canada. The approach of this paper has been inspired by the work of Alan Manne and I thank Alan for his advice in developing the original model on which the computational model of this paper is based. All opinions and any errors are my responsibility alone.

[1] For natural gas recovery, Kalisch and McGill (1990, 2) write: "...the rate of technological advancement has increased so rapidly in recent years that resources inaccessible twenty years ago, are now routinely being developed and produced.

[2] Brief references to technological change are scattered throughout. Yet chapters on minerals resource stocks and information, strategies for modeling exhaustible resource supply, and energy, the environment and economic growth, contain virtually no discussion of technological change.

[3] Of the studies cited, two are closest to this work. Chermak and Patrick (1995) utilize an exhaustible resource framework to examine the benefits of an enhanced information technology and focus on a single gas well, with prices exogenous. Rogner (1988) examines issues that are somewhat similar, but in an energy model encompassing oil, coal, natural gas, nuclear and renewables. He discusses a conventional technology scenario and a technical evolution scenario but provides no welfare comparisons.

[4] Rogner (1988), Lohrenz (1991) and Chermak and Patrick (1995) provide useful discussion of some of the technical aspects of gas production.

[5] A third nonconventional source is gas hydrates. Extremely large amounts of hydrates are believed to exist, but commercializability is a fantasy.

[6] Rowse (1996) provides detailed discussion of the model, which employs several modifications of earlier versions used by Rowse (1986, 1990).

[7] Gas market deregulation during the past decade has profoundly changed gas markets. National Energy Board (1993) argues that North American markets are integrating; hence assuming exogenous export prices becomes questionable. A more defensible approach *might* be to utilize a dynamic multiregional North American model which simultaneously solves for prices in all regions and periods. But no such model was available and constructing such a model was prohibitive. Moreover, because markets are still evolving, the results of *any* integrated model may be questionable.

[8] A recovery profile also applies to deliverable gas from proved reserves which is not produced. Again, 10% annual decline is assumed.

[9] Computing user costs is central to identifying efficient allocations. Using an earlier model, Rowse (1986) shows that user costs are computed when the Kuhn-Tucker optimality conditions are satisfied.

[10] This stark behaviour of exports is not implausible. BC supplies compete with US supplies in the US, and when the BC price rises above the corresponding export price, exports collapse. In reality it is unlikely that the export market would be abandoned abruptly, but a threshold price could exist at which US supplies would largely displace BC exports.

[11] Given equation (A7) below for Q_t, the integral under each domestic demand function -- measuring gross consumer surplus -- does not converge. Rowse (1990, 778) discusses how a choke price solves this problem. Of course, with technology changing, a choke price may be a moving target.

[12] Domestic consumption and exports sum to slightly less than the aggregate production from conventional and backstop sources because some gas is used for intraprovincial pipeline fuel.

[13] Rogner (1988, 13) writes: "... how one can introduce technological change into natural gas studies without drifting off into science fiction ..."

[14] How long can technological progress go unrecognized? One decade is used but this duration may be too long; other choices are possible.

[15] Comparing SW of $37.470 B for Case 1C with SW of $33.501 B for 1B allows ignorance of the future to generate a large welfare gain over perfect foresight! But there is no paradox: due to revised knowledge at the end of 1999 in Case 1C, there is more inertia in the 1C demands for 2000-2079 than 1B, and larger welfare simply reflects greater demand inflexibility.

[16] Whether this reduction is too little or too much is unknown, but it does speed the introduction of new conventional supplies, as Table 1 reveals.

[17] ξ_3 becomes -0.122 and ξ_2 becomes negative, giving h(W) an asymptote.

[18] For instance, technological breakthroughs could make large volumes of nonconventional gas available at much reduced cost, but pipeline and distribution costs would not change, and the price to consumers might not change much. Moreover, slow changes to the gas-using capital stock mean that gas use would not change much until several or many years pass. Further, to cope with financing requirements and risks of large-scale production, producers would increase nonconventional supplies gradually over time. These actions shrink the discounted benefits of technological advances. Burgeoning exports could undercut these arguments somewhat.

[19] Only paramters a_1.through a_{30} are nonzero because the recovery profile terminates after three decades. But defining $a_{31} = ... = a_T = 0$ yields economy in notation. The same holds true for σ_{31} through σ_T below.

[20] See note 11.

[21] Salvage value assumes that backstop gas determines the domestic price by the horizon, and thus the domestic price must migrate endogenously to the delivered cost of backstop gas or salvage value will be measured wrongly. Salvage value consists of the wellhead gas value less production costs, all multiplied (and discounted) by flows of gas producible after 2064.

Chapter 3

A CONCURRENT AUCTION MODEL FOR TRANSMISSION CONGESTION CONTRACTS

William W. Hogan[1]

Harvard University

Introduction

Network interactions in the electricity system create externalities that have precluded the development of a workable system of fully decentralized "physical" property rights for controlling use of the transmission system in an open access, competitive electricity market. [2] Transmission congestion contracts provide a well-defined alternative mechanism to serve in the place of strictly physical property rights related to transmission usage. Any of a number of methods could provide an initial allocation of transmission congestion contracts. For instance, existing users might receive a designated set of contracts based on historical usage patterns, and then the remainder could be assigned to new users. With the availability of well defined transmission congestion contracts, it would be natural to employ an auction for allocating part or all of the contracts to allow for non-discriminatory access through a market mechanism.

The details of an auction could accommodate many special features of the transmission system. The essence in the context of a contract network framework is to ensure that the allocated contracts are feasible given the configuration of the network. In this case, a straightforward adaptation of an optimal power flow dispatch model provides a formulation of a concurrent auction model for selecting the long-term capacity awards based on the willingness to pay. The power flow formulation of the auction avoids the necessity of identifying which transmission congestion contracts are available by characterizing all possible contracts and selecting the combination of feasible contracts that would provide the highest valued use of the network.

Transmission Congestion Contracts

The context for the creation of transmission congestion contracts (TCC) is a system of short-term transmission usage pricing based on locational marginal cost pricing.

John Weyant (ed.), ENERGY AND ENVIRONMENTAL POLICY MODELING. Copyright © 1998. Kluwer Academic Publishers. ISBN 0-7923-8348-6. All rights reserved.

An independent system operator (ISO) determines the locational prices based on the actual dispatch and the bids of system users, and either buys and sells power at these prices or charges the locational differences in these prices for transmission of power from one location to another. For most transmission, large differences in locational prices would be dominated by the difference in congestion costs. All transmission usage would be charged based on these locational differences.

A general description of a TCC could be any vector of net loads in the grid. The typical discussion of TCCs presumes that the vector describes transmission of a fixed amount of power from a source to a destination in the network. This special case for transmission of "x" MW would be the vector

$$\text{TCC} = (0,..,\ -x_{source},\ ...,\ x_{destination},\ ...,0)^t.$$

This form of a balanced TCC always sums to zero. However, there would be no necessity to impose this balancing requirement on each individual TCC. All that would be required would be that the set of all TCCs would be simultaneously feasible and appropriately balanced.

The TCC would be denominated in the quantity of power input and output at various locations. This is similar to transmission from source to destination, intended to mean the actual flow of power, or at least specific performance on the locational delivery of the power. However, the TCC is not a contract for actual delivery of specific, identified power. The definition assumes that loads will be met either through actual delivery or through displacement. Hence, the actual power flows may be (very) different from the quantities embodied in the collection of TCCs. By contrast to a contract for physical flows, the TCC is a contract for payment of congestion costs. These payments are designed so that the user is economically indifferent between meeting the load through actual delivery or through displacement. [3]

The short-term locational prices for the actual dispatch can be decomposed into three components relative to a reference bus: the price of generation at the reference bus, the marginal cost of losses relative to the reference bus, and the marginal cost of congestion relative to the reference bus. Let p_C be the vector of congestion prices for each bus. The contract between the ISO and the holder of TCC_i calls for a payment by the ISO of $p_C^t(\text{TCC}_i)$. For a balanced TCC_i from a location with a low congestion price to a region with a high congestion price, the TCC_i payment to the TCC_i holders would be positive, just compensating for the congestion cost of the price of transmission usage. In the reverse case, the TCC_i holder would make a payment to the ISO, returning the negative transmission usage charge paid by the ISO. [4] Hence, the TCC would not affect the dispatch or give the holder any control over the use of the transmission grid. However, in each case the holder of the TCC could perfectly hedge the congestion cost of transmission usage as though power had flowed according to the TCC but free of congestion cost.

In this sense, a balanced TCC is analogous to a futures contract for the spot price of transmission congestion, with a target price of zero.   If the spot price of transmission congestion were more (or less) than zero, the TCC would exactly balance the spot price payment for the quantity covered by the contract. This TCC could be traded in a secondary market and would provide a contractual mechanism for long-term pricing of transmission in a competitive, open access electricity market.

Concurrent Auction

Consider first the case of a model of real power only, as in the DC-Load approximation. For this simplified case, define the net real power loads at each bus as the vector y_P of load minus generation. [5] The possible set of real power loads is constrained by the usual network load flow equations and a series of constraints. These constraints could include MW limitations on line flows in both pre- and post-contingency conditions.   There may be many of these constraints, including complicated limitations involving multiple lines and locations.   The only requirement is that the constraints can be specified in terms of the net loads. [6] For the present discussion, all of these constraints are collected in the function K(.), with the feasible net loads characterized by:

$$K(y_P) \leq 0.$$

If congestion payment obligations must be met from the congestion revenues collected by the ISO, the TCCs must be feasible, in the sense that $TCC = \Sigma TCC_i$ and $K(TCC) \leq 0$. That feasibility would be necessary is clear from the case where the only transmission usage is from the net loads implied by the TCCs. In the DC-Load case, where all the constraints K(.) are linear, feasibility is also sufficient to assure this revenue adequacy condition. If the actual usage of the system is y_P^*, and TCC is feasible, then we know that

$$p_C^t(y_P^*) \geq p_C^t(TCC) \quad .$$

In other words, the congestion payments collected by the ISO for actual use of the system would always be at least as large as the congestion payments made to the holders of the TCCs. [7] Hence, the ISO would always be hedged. [8] As long as the grid is the same, and only load patterns are changing, the ISO would be able to honor the TCC commitments. [9]

Presumably the use of a TCC would be as a hedge for the congestion component of transmission costs for a long-term power sale. The price that users would be willing to pay for a TCC would be limited by the economics of the power deal. Suppose that this or some other economic assessment permits users to evaluate TCCs.   This is equivalent to asking the user to set a value for long-term transmission between locations. Following the motivation of competitive markets, an auction would provide an equilibrium allocation of the TCCs at market prices, with the assignments to the highest valued uses.

The infinite array of possible feasible TCCs makes it difficult to define in advance the availability of any subset of the contracts; all the TCCs would interact in the network, and the many TCCs would be separate products that must be auctioned concurrently. [10] However, it is possible to structure a concurrent auction that simultaneously defines and awards the TCCs. A description of the network and its constraints provides a characterization of all possible combinations of TCCs. Following the logic of economic dispatch, the market equilibrium for these multiple products will be equivalent to the result of a central evaluation of bids for the TCCs under the assumption that the participants have an incentive to bid their maximum willingness to pay. If all winning bidders would pay the market clearing price for their TCCs, and there were enough bidders so that no bidder would know in advance which bid would set the market price, then the participants would have an incentive to bid their maximum willingness to pay, and the centralized concurrent auction of TCCs would achieve the market equilibrium.

The resulting TCC concurrent auction optimization problem would be closely related to the corresponding economic dispatch problem. Suppose that we describe a bid for capacity by bidder "i" as a maximum quantity $TCCBID_i$, with vector bid_{Pi} that defines the real power inputs and outputs per unit of the TCC. The accompanying maximum price would be $Pbid_i$. Let x_i be the allocation of TCC_i. Then under the DC-Load assumption of ignoring losses, the adaptation of the optimal allocation of TCCs problem becomes:

$$\text{Max} \qquad\qquad \Sigma\, Pbid_i x_i$$
$$x_{I \geq 0}, y_P$$

subject to   Bid Definition

$$x_i \leq TCCBID_i, \text{ for all } i,$$
$$y_P - \Sigma\, bid_{Pi} x_i = 0 \;\; ;$$

Kirchoff's Laws and System Operating Limits
$$K(y_P) \leq\; 0\; .$$

The solution to this problem will yield the optimal TCC awards. Furthermore, under the assumption that the bids represent the maximum willingness to pay, the dual solution yields the market clearing prices for the bids. [11] In particular, the corresponding dual variables and optimality conditions would include:

$$\theta_i + (\lambda_P)^t bid_{Pi} \geq\; Pbid_i,$$
$$\lambda_P - \nabla K \mu = 0\; ,$$
$$\theta_i,\; \mu \geq 0\; .$$

Apparently the opportunity cost of congestion for each location would be defined by the vector λ_P, and the opportunity cost for a particular award of TCC_i would be $(\lambda_P)^t bid_{Pi}$. By the principles of complementary slackness, for any positive award of a TCC, the marginal opportunity cost price would be $P_{TCCi} = (\lambda_P)^t bid_{Pi}$ and the bidder's surplus or rent would be $\theta_i = Pbid_i - P_{TCCi}$. Since this surplus is always positive, we see that P_{TCCi}, which would be the market clearing TCC award price paid, would never be greater than the bid price. [12]

Example Auction

The three bus example network in Figure 1 illustrates the elements of a concurrent auction of TCCs. Here the three buses are connected by three identical lines. We follow the DC-Load assumption and ignore losses. There is only one constraint which limits the flow of power on the line between buses 1 and 3 to a maximum of 600 MW. The various actors in the market have identified two types of TCCs that would have value, from bus 1 to bus 3 and from bus 2 to bus 3. The assumption is that there are many bidders with different maximum evaluations of the amount they would pay for the respective TCCs. These evaluations become bids in the concurrent auction. The collection of all the bids appears as a bid curve for each type of TCC. For simplicity, the bids are assumed to be the same for both types of TCCs, but any bids would be allowed.

The three bus example is the simplest case that include s the effects of loop flow and network interactions. However, there is no necessary connection between the definition of the TCCs and the ownership of the lines between buses. The example could be expanded by adding other lines and buses. The TCCs would still be defined from one bus and to another bus, without any requirement that there be a direct link between the two buses.

Here the highest bid is at 6 cents, and the bid prices decline to zero at the level of 1200 MW. The objective is to find the combination of awards that maximizes the area under the bid curves, which is the sum of the value of the successful bids. In principle, all the transmission capacity could be awarded to TCCs from either source. If all the TCCs came from bus 1, then the line limit would constrain and the maximum award possible would be 900 MW with a price of 1.5 cents. The value would be the area under the bid curve, $1.5(900) + 4.5(900)/2 = 3375$. If all 1200 MW of bids for TCCs from bus 2 were accepted, the price for these would be zero and there would be excess capacity. The value for these awards would be the area under the bid curve, $0(1200) + 6(1200)/2 = 3600$. Neither extreme would provide the highest valued use of the transmission grid. However, the concurrent auction formulation takes into consideration all the bids and the interactions in the network to find the maximum value award and the associated market clearing prices for the TCCs.

7

Figure 1

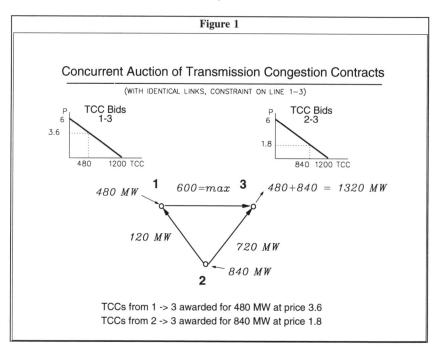

Concurrent Auction of Transmission Congestion Contracts

(WITH IDENTICAL LINKS, CONSTRAINT ON LINE 1–3)

TCCs from 1 -> 3 awarded for 480 MW at price 3.6
TCCs from 2 -> 3 awarded for 840 MW at price 1.8

The result of the concurrent auction in Figure 1 awards 480 MW for TCCs from bus 1 to bus 3 and 840 MW of TCCs from bus 2 to bus 3. The market clearing prices for the respective TCCs are 3.6 cents and 1.8 cents. The value for these awards would be the area under the bid curves, $3.6(480) + 2.4(480)/2 + 1.8(840) + 4.2(840)/2 = 5580$. In this simple case, the ratio of the prices is just the inverse of the tradeoff between the two types of TCCs. In order to maintain feasibility, given the constraint on the line from 1 to 3, each MW from bus 1 to bus 3 displaces 2 MW from bus 2 to bus 3.[13]

The existence of the TCCs tells us nothing about the total price of power that might be arranged under contract or that would be determined in the spot market. Apparently the winning TCC bidders believe the average differences in the prices between buses will be at least as large as the concurrent auction award prices. However, with these TCCs in place, the holders would have a perfect hedge for the spot price of transmission. If the spot price of transmission is high, then the TCC congestion payment would compensate the holder for the spot price. However, the spot price of transmission could be higher or lower than the cost of the TCC.

For example, suppose that the actual dispatch conditions conform to those in Figure 2 where economic dispatch leads too much of the load at bus 3 being supplied by generation at bus 3 with a cost of 2.6 cents. Here the generation at bus 2, where the opportunity cost is 2.3 cents, is too expensive to run, and the remaining generation at bus 1 is supplied at a price of 2 cents.

8

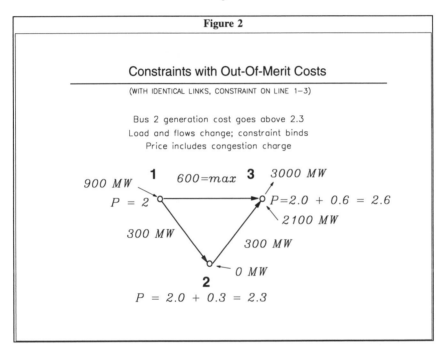

Figure 2

Constraints with Out-Of-Merit Costs

(WITH IDENTICAL LINKS, CONSTRAINT ON LINE 1–3)

Bus 2 generation cost goes above 2.3
Load and flows change; constraint binds
Price includes congestion charge

1 600=max **3** 3000 MW

900 MW

$P = 2$ $P = 2.0 + 0.6 = 2.6$

2100 MW

300 MW   300 MW

0 MW

2

$P = 2.0 + 0.3 = 2.3$

Everyone using the transmission grid is paying at these short-term prices. Those buying and selling through the ISO employ the appropriate locational prices. Those transmitting power from one bus to another pay the spot price of transmission equal to the difference in the locational prices. The total net usage charge collected by the ISO is 3000(2.6) - 2100(2.6) - 900(2) = 540. The difference in congestion charges between bus 1 and bus 3 is 0.6, requiring a payment of 480(0.6) to the holders of TCCs from bus 1 to bus 3. The difference in congestion charge from bus 2 to bus 3 is 0.3, requiring a payment of 840(0.3) to the holders of TCCs from bus 2 to bus 3. The total payment to TCC holders is 480(0.6) + 840(0.3) = 540. Hence, the total congestion payments for use of the grid are large enough to pay the TCC obligations, even though the dispatch and economic conditions have changed.

For the holders of TCCs from bus 1 to bus 3, the effective cost of power delivered at bus 3 is 2 cents, the same as the price at bus 1; for the holders of TCCs from bus 2 to bus 3, the effective cost of power delivered to bus 3 is 2.3 cents, the same as the price at bus 2. The holders of the TCCs have a perfect hedge for the spot price of transmission congestion. Of course, in this case the holders paid more for the TCCs than they were worth for this particular dispatch. With economic conditions changing, presumably there would be other periods when congestion could be greater than the price paid for the TCC. Whether the average congestion costs would justify the price of the long-term protection is uncertain, and would be a business risk in the competitive market. However, the TCC holders would be assured of getting what they paid for: long-term protection from the uncertain congestion costs of transmission, no matter what the changing pattern of the loads.

Including Losses and Reactive Power

The generalization of the TCC auction to account for the effects of losses, reactive power and the non-linear AC-load formulation follows in a natural way. Here the TCC would be defined in terms of both real and reactive power net loads. However, the model must be expanded to account for losses. Because the TCC calls only for payment of congestion rents, which are by definition set to zero at the reference bus, the losses or other imbalances can be treated as being met at the reference or swing bus, as the quantities y_{Ps} and y_{Qs} needed to satisfy the power balance equations. [14]

The constraints on the vector net real loads y_P and reactive loads y_Q would now become $K(y_P, y_Q, y_{Ps}, y_{Qs})$ and include pre- and post-contingency MVA line flow limits as well as bus voltage magnitude bounds. Under certain assumptions for this more general case, a similar revenue adequacy condition would apply: for any feasible set of TCCs and any actual dispatch, the short-term congestion payment obligations under the TCCs would always be no more than the congestion revenues collected by the ISO. [15] Hence, maintaining a feasible set of TCCs would be necessary for this riskless hedge, and under market equilibrium conditions would be sufficient. The objective of the auction is to find the highest valued allocation of the feasible TCCs.

The corresponding non-linear concurrent auction problem would become:

Max $\Sigma\ Pbid_i x_i$

$x_I \geq 0,\ y_P,\ y_Q,\ y_{Ps},\ y_{Qs}$

subject to   Bid Definition

$$x_i \leq TCCBID_i, \text{ for all } i,$$
$$y_P - \Sigma\ bid_{Pi} x_i = 0\ ,$$
$$y_Q - \Sigma\ bid_{Qi} x_i = 0\ ;$$

Kirchoff's Laws and System Operating Limits

$$K(y_P, y_Q, y_{Ps}, y_{Qs}) \leq 0 \ .$$

As before, the dual constraints and variables would define the market clearing prices. The dual problem includes:

$$\theta_i + (\lambda_P)^t bid_{Pi} + (\lambda_Q)^t bid_{Qi} \geq Pbid_i,$$

$$\lambda_P - \nabla K_P \mu = 0 \ ,$$
$$\lambda_Q - \nabla K_Q \mu = 0 \ ,$$

$$\theta_i, \mu \geq 0 \ .$$

Now the market value of congestion for each location would be defined by the vectors λ_P and λ_Q, and the opportunity cost for a particular award of TCC_i would be $(\lambda_P)^t bid_{Pi} + (\lambda_Q)^t bid_{Qi}$. By the principles of complementary slackness, for any positive award of a TCC, the marginal opportunity cost price would be $P_{TCCi} = (\lambda_P)^t bid_{Pi} + (\lambda_Q)^t bid_{Qi}$ and the bidder's surplus or rent would be $\theta_i = Pbid_i - P_{TCCi}$. Since this surplus is always positive, the market clearing TCC award price paid would never be greater than the bid price.

Extensions

The TCC concurrent auction optimization problem is essentially a special case of the non-linear optimal power flow dispatch problem. The complication is only in the addition of a few linear side constraints -- the "Bid Definition" constraints -- which would be easy to implement. Although optimal power flow models may be difficult to use in real time for fully automated control of the system, this auction calculation need be repeated only at infrequent intervals when there is to be an expansion of the system or trading among the existing TCCs.

Further extensions of the complexity of the auction could be accommodated. For example, bidders may submit multiple bids, and then apply constraints on the joint awards across these bids. Any set of added linear constraints on the bidders set of x_i variables should be easy to incorporate, with more non-linear constraints depending on the availability of software to solve the problem. Presumably any bidder's constraints on its own bids could be accommodated, as long as zero (i.e., no award) would be a feasible solution to the set of side constraints.

The formulation of the concurrent auction model is quite general, but it presumes the ability of the bidders to define their preferences. In the case of real power flows, losses would be small and it would be reasonable to expect the bidders either to define imbalanced bids to account for losses or to accept the losses computed as necessary to balance at the reference bus. For reactive power, however, balancing by individual TCCs is less reasonable and the different levels of reactive power needed to support a TCC would be more difficult to determine. Here an approximation may be obtained by incorporating a range of reactive net loads that would accompany each TCC bid. Formally this could be accomplished through multiple bids, with constraints across bids. Or a single bid could include maximum and minimum levels of reactive net input associated with a real power bid.

The concurrent auction could be implemented as a single-pass system or as part of a sequential auction. The concurrent auction deals with the interactions in the network, but not with interactions with generation and load contracts that might be relevant in determining the bid maximum prices, $Pbid_i$. In a sequential version of the concurrent auction, the bids could be revised for a sequential repetition of the auction for a fixed number of cycles or until no bids changed. Subject to certain limitations on the bid changes to avoid strategic behavior and cycling, this could provide the auction participants with additional market information in setting their maximum bid prices. [16]

Note that this auction model allows for consideration of existing TCCs. These can be included simply as bids, with the minimum selling price treated as the bid. If the existing TCC is not included in the award, then it has been sold back to the market. The sales price is still P_{TCCi}, but for variables not included in the optimal solution the principle of complementary slackness guarantees that this sales price would be at least as high as the bid price. Hence, for TCCs which the holders do not wish to sell at any price, a sufficiently high minimum selling price will guarantee the TCC is preserved, and the holder of the TCC would formally sell and buy the TCC at the market price, leaving no change.

The concurrent auction formulated above is for a single, static set of TCCs. This model could be applied to separate periods to allow for time varying TCCs, such as peak and off-peak. In the UK, for example, power contracts are written that can differ in prices and terms for each of the 8760/4=2190 four hour long "electric forward agreement" (EFA) periods each year. In principle, the concurrent auction model could be extended to include multiple periods with inter-temporal constraints. For instance, bids may be for multiple periods, with different bids covering different periods. The network and inter-temporal constraints could assure feasibility, with the objective function specified as the present value of the bids. For the DC-Load formulation, even for large networks this optimization problem would be no more complicated than the dispatch problems now solved routinely for all 8760 hours of the year. For the full non-linear problem, current software can solve a single period optimization in a few minutes for a large network. The extension to multiple periods is possible, but remains to be demonstrated.

Conclusion

The complex interactions in real electric networks make it impossible to define the capacity for point-to-point reservations or transmission congestion contracts in the simple additive manner that would be possible for a radial system. However, it is possible to characterize the constraints on the feasible set of transmission flows, and this is done as a regular part of economic dispatch. The extension to operate a spot-market using a bid-based economic dispatch suggests the parallel extension to a concurrent auction for awarding a feasible set of transmission congestion contracts. The same optimal dispatch formulation, extended to include a set of linear side constraints to define the bids, allows bidders to express preferences for transmission congestion contracts and then the ISO to determine the simultaneous set of awards and market clearing prices that maximize the value of the awards as expressed in the bids. Hence the auction can implicitly include all possible awards, without any necessity of forecasting a particular set of loads or transmission congestion contracts.

Endnotes

[1] William W. Hogan is the Thornton Bradshaw Professor of Public Policy and Management, John F. Kennedy School of Government, Harvard University, and Director, Putnam, Hayes & Bartlett, Inc., Cambridge MA. He serves as Research Director for the Harvard Electricity Policy Group. The author is or has been a consultant on electric market reform and transmission issues for British National Grid Company, General Public Utilities Corporation (working with the "supporting" companies of the PJM proposal), Duquesne Light Company, Electricity Corporation of New Zealand, National Independent Energy Producers, New York Power Pool, New York Utilities Collaborative, San Diego Gas & Electric Corp., Trans Power of New Zealand, and Wisconsin Electric Power Company. Hamish Fraser, Scott Harvey, Laurence Kirsch, Richard O'Neill and Susan Pope provided helpful comments and assistance. The views presented here are not necessarily attributable to any of those mentioned, and the remaining errors are solely the responsibility of the author.

[2] W. Hogan "Electricity Transmission Policy and Promoting Wholesale Competition," Initial Response to the Notice of Proposed Rulemaking Regarding Promoting Wholesale Competition Through Open-Access Non-Discriminatory Transmission Services by Public Utilities, Federal Energy Regulatory Commission, Docket No. RM95-8-000, Harvard University, August 7, 1995.

[3] There is much less to the distinction between physical and financial transmission rights than meets the eye. In the narrowest definition of strictly physical rights, no use of the transmission system would be allowed without obtaining in advance a matching "physical" reservation that would be acquired somehow in the initial allocation or the secondary market. But practical implementations provide that the reservations could be used either in the "physical" sense of matching actual transmission flows to reservations, or in a "financial" sense in that unused reservations would in effect be bought, sold and reconfigured based on opportunity cost pricing through a spot market coordinated through a bid-based economic

dispatch by the ISO. In the presence of an active spot market, this trading of capacity reservations at opportunity cost prices makes the "physical" capacity reservation more like a "financial" contract. Whether we call these transmission instruments "capacity reservations" or "transmission congestion contracts" or something else is more a semantic than a substantive issue. How we organize the market, however, does matter because of the rather substantial transaction costs. Forcing people to think of, and treat, the transmission capacity reservations as narrowly defined physical rights that have to be reconfigured, constantly and explicitly, in order to allow for schedules in the transmission system, would create a cumbersome and possibly unworkable system for the actual dispatch. Recognizing that economic dispatch reconfigures and trades these rights implicitly would capitalize on well-established principles of reliable dispatch and economic efficiency. For further details, see S. M. Harvey, W. W. Hogan and S. L. Pope, "Transmission Capacity Reservations and Transmission Congestion Contracts," Harvard University, June 6, 1996, (revised October 14, 1996 and filed with the FERC as part of submission of William W. Hogan, Capacity Reservation Open Access Transmission Tariffs Response to FERC Notice of Proposed Rulemaking, Docket No. RM96-11-000, Washington, D.C., October 21, 1996).

[4]This is the "obligation" form of the TCC. If the payments were discretionary, then the holder would never make a payment to the ISO for the case of a negative congestion difference. This would be the "option" form of the TCC. The option form would require a substantially more complicated feasibility test, and is not considered here. For further discussion, see S. M. Harvey, W. W. Hogan and S. L. Pope, "Transmission Capacity Reservations and Transmission Congestion Contracts," Harvard University, June 6, 1996, (revised October 14, 1996).

[5]If the demand for power at each location or bus is "d" and the generation is "g", let y= d-g be the vector of net loads at each bus. The sign convention reverses the approach in Schweppe et al., but simplifies the interpretation of prices. F. C. Schweppe, M. C. Caramanis, R. D. Tabors, and R.E. Bohn, *Spot Pricing of Electricity*, Kluwer Academic Publishers, Norwell, MA, 1988.

[6]The distinction refers to the common practice of using "interface" constraints limiting the power flow on particular lines. The "limit" often is a projection assuming small deviations from a target value of net loads, say t_P. In effect, the constraint can be interpreted as being an approximation of the form of $K(y_P, t_P) \leq 0$, with t_P fixed and y_P close to t_P. If y_P differs from t_P, then the "limit" is changed and a new target is set. For our purposes, we would interpret the constraint as $K(y_P) = K(y_P, y_P) \leq 0$.

[7]W. Hogan, "Contract Networks for Electric Power Transmission," *Journal of Regulatory Economics*, Vol. 4, September 1992.

[8] In principle, speculators could offer any amount of TCCs, for a price, but the ISO would not be required to accept any exposure.

[9]There may be excess congestion rentals after making all the required payments under the TCCs. These payments should not be left with the ISO and could be redistributed under a sharing formula, perhaps among the owners of the TCCs.

[10]W. Hogan, "An Efficient Concurrent Auction Model for Firm Natural Gas Transportation Capacity," *Information Systems and Operational Research*, Vol. 30, No. 3, August 1992.

[11]R. P. O'Neill and W. R. Stewart, Jr., "A Linear Programming Approach for Determining Efficient Rates for Public Utility Services," *Advances in Mathematical Programming and Financial Planning*, Volume 3, JAI Press, 1993, pp. 163-178.

[12]Note that either the maximum price bids, $Pbid_i$, or the TCC award prices could be negative, implying that the holder would be paid in advance to take on the financial obligation for transmission flow that apparently would increase the overall capacity of the system. For example, a TCC might provide counterflow in parts of the network that increased the capability to award other TCCs.

[13] The bid vectors are $(-1, 0, 1)^t$ for the TCC from bus 1 to bus 3 and $(0, -1, 1)^t$ from bus 2 to bus 3. The values for the various dual components include: $\nabla K = (0, 1/3, 2/3)^t$; $\mu = 5.4$; $\lambda_P = (0, 1.8, 3.6)^t$.

[14]The convention to apply the TCC idea only to congestion is not required. Losses could be included, at the cost of requiring some unbalanced TCCs. See S. M. Harvey, W. W. Hogan and S. L. Pope, "Transmission Capacity Reservations and Transmission Congestion Contracts," Harvard University, June 6, 1996, (revised October 14, 1996).

[15]Whenever there is a spot-market equilibrium set of prices, the revenues from the actual dispatch will exceed the payments required under the TCCs for the general case including losses. In general, there may be conditions where there is no set of equilibrium prices available, due to non-convexities in the optimal dispatch problem. In this (probably rare) case, central dispatch would be required to achieve a welfare maximizing solution, but the ISO might not be able to guarantee the full revenue for the TCCs. For further discussion, see S. M. Harvey, W. W. Hogan and S. L. Pope, "Transmission Capacity Reservations and Transmission Congestion Contracts," Harvard University, June 6, 1996, (revised October 14, 1996).

[16]This has been suggested, for example, in the New York Power Pool proposal. See "Responses to Questions Regarding the Report on NYPP Proposed Market Structure," New York Public Service Commission filing by the New York Power Pool, January 3, 1997, questions NYPP-103; PSC-19 through NYPP-108; PSC-24.

Chapter 4

SPOT MARKETS IN ELECTRIC POWER NETWORK: THEORY
Hung-po Chao and Stephen Peck[1]

1. Introduction

The movement toward market competition in the U.S. began in the 1970's with some capital intensive network-based industries such as transportation, energy, and telecommunications. The potential for increased competition in the electric power industry has long been discussed. Back when interconnected systems took shape after the innovation of high voltage transmission, the electric power system exhibited notable characteristics of natural monopoly. In general, as a network system becomes bigger and better connected, the number of potential options available to customers increases, and the pressure for greater reliance on market mechanisms grows stronger. In the 1990's, as competitive forces sweep across the electric power industry around the world, access and pricing policy for transmission will play a pivotal role in shaping future market structure and performance. However, pricing for the electric power transmission system is complicated by some unique characteristics of the electric power network. An electric power grid differs technologically from other types of network in that power flows must observe Kirchoff's laws. This gives rise to the `loop flow' phenomenon, creating widespread externalities in the markets for electric power, whose complexity only grows with the size of the system. Therefore, it is commonly assumed that horizontal integration of transmission is necessary to coordinate resource allocation efficiently. Joskow and Schmalensee (1983) point out the fundamental importance of the externality problem in evaluating alternative proposals for restructuring. Hogan (1992) discusses some of the significant difficulties caused by the externalities in restructuring the electric power transmission grid.

Chao and Peck (1995) present a new approach to the design of an efficient market mechanism that incorporates the externalities due to loop flows in the electric power network. The primary purpose of this paper is to present a simplified model so that the main ideas can be more accessible to readers who are not as familiar with the technical complexity of a power system. For this purpose, we use a

John Weyant (ed.), ENERGY AND ENVIRONMENTAL POLICY MODELING. Copyright © 1998. Kluwer Academic Publishers. ISBN 0-7923-8348-6. All rights reserved.

linear DC load flow model of electric power and ignore transmission losses. The essential elements of the new approach include a set of tradable transmission capacity rights and a trading rule that specifies the transmission capacity rights required for each power transaction. The trading rule makes explicit the effects of an electric power transaction on the power flows throughout the network. Therefore, the trading rule enables the opportunity cost for utilizing the network to be incorporated into the decisions involved in each transaction. Under this scheme, we demonstrate that a competitive equilibrium is a social optimum and that a dynamic maket process with electricity trading and transmission bidding is Lyapunov stable and always converges to a competitive equilibrium. Overall, the new approach provides a conceptual vision for the design of decentralized markets for electricity.

The organization of the paper is as follows: In Section 2, we present a simple model of an electric power network; we characterize a linear approximation to power flow equations and the social welfare function. In Section 3, we present the design of a market for transmission capacity rights and demonstrate its efficiency properties. In Section 4, we characterize a dynamic trading process and investigate the dynamic behavior of the mechanism. In Section 5, we present a numerical example that illustrates the convergence of the dynamic trading process. In Section 6, we conclude with a brief summary of key results and remaining issues for future research. We provide the proofs for all the propositions in the Appendix.

2. A Simple Model of Electric Power Network

The electric power network is less flexible than other network systems. In a transportation network, a buyer and a seller at two different locations can choose alternative routes to deliver merchandise. Once a route is chosen then a contract may be written which specifies the unique physical transport path. In other words, the contract path and the physical transport path coincide. In an electric power network, although contracts are frequently written assuming a unique physical path, the contract path is usually different from the physical path, since power flows must observe Kirchoff's laws, and the physical path tends to ``diffuse" out from the contract path in a way that minimizes the losses. The system controls are mainly limited to the dispatch of power generation and load management. In this section, we consider a simplified model of an electric power network for investigation of the economic implications of Kirchoff's laws.

Power flow equations

We consider an electric power network with n nodes that supports alternating currents. The alternating voltage at node i is represented by a sinusoidal wave form with amplitude, V_i, and voltage angle relative to the voltage at a reference node, θ_i. The power flows can be divided into real power and reactive power. For our purposes, we shall focus on the real power flows. From Kirchoff's laws, the real power flow equations can be written (Elgerd 1982, Graves 1995):

$$P_{ij} = G_{ij}V_i^2 - G_{ij}V_iV_j\cos(\theta_i - \theta_j) + Y_{ij}V_iV_j\sin(\theta_i - \theta_j) \quad (1)$$

and

$$q_i = q_i^s - q_i^d = \sum_{j=1}^{n} P_{ij}, \qquad i = 1, \cdots, n \qquad (2)$$

where $Y_{ij} \equiv x_{ij} / (r_{ij}^2 + x_{ij}^2)$ and $G_{ij} \equiv r_{ij} / (r_{ij}^2 + x_{ij}^2)$; r_{ij} and x_{ij} are, respectively, the resistance and the inductance of line (i,j); P_{ij} is the real power flow from node i to node j; q_i^s is the power generation at node i; q_i^d is the power demand at node i; and q_i is the net power supply from node i. By convention, when P_{ij} is negative, the power flows from j to i.

In this paper, we make the following simplifying assumptions: 1) $G_{ij} = 0$, for $1 \leq i,j \leq n$, which means that the line resistances are zero, and thus there are no transmission losses in the network, (generally, the line resistance is much less than the line inductance.) 2) $V_i = 1$, for $1 \leq i \leq n$, i.e., the voltages are fixed at rated levels, (reactive power is freely available) and 3) $\theta_i - \theta_j \approx 0$, i.e., the power angle differences are small. As a consequence of these assumptions, the power flow equations in (1) - (2) can be aproximated by the following linear DC load flow equations:

$$P_{ij} = Y_{ij}(\theta_i - \theta_j), \quad \text{for } 1 \leq i, j \leq n, \qquad (3)$$

and

$$q_i = \sum_{j=1}^{n} Y_{ij}(\theta_i - \theta_j) \qquad \text{for } i = 1, \dots n. \qquad (4)$$

Social welfare function

We assume that the value of power consumption can be represented by a nondecreasing concave benefit function, $B_i(q_i^d)$, and the minimum cost of power generation can be represented by a nondecreasing convex cost function, $C_i(q_i^s)$. We define below the social welfare at node i as a function of net electricity consumption $\bar{q}_i \equiv -q_i = q_i^d - q_i^s$:

$$W_i(\bar{q}_i) \equiv \max_{q_i^d, q_i^s} B_i(q_i^d) - C_i(q_i^s), \qquad (5)$$

subject to

$$\bar{q}_i = q_i^d - q_i^s. \qquad (6)$$

Note that $W_i'(\bar{q}_i) = B_i'(q_i^d) = C_i'(q_i^s)$ when q_i^d and q_i^s are set optimally. The social welfare maximization problem is stated as follows:

$$\max_{\theta,\hat{q}} \sum_{i=1}^{n} W_i(\overline{q}_i)$$

(7)

Subject to

$$\overline{q}_i = -\sum_{j=1}^{n} Y_{ij}(\theta_i - \theta_j), \qquad \text{for } i = 1,...,n$$

(8)

$$Y_{ij}(\theta_i - \theta_j) \le \overline{P}_{ij}, \qquad \text{for } 1 \le i, j, \le n$$

(9)

where $\overline{P}_{ij}$ is the capacity of line (i,j), and (9) represents the transmission capacity constraint.

Since the power flow equations in (8) depend only on the difference of voltage angles, there are n-1 independent voltage angles. Without loss of generality, we assume that $\theta_n = 0$. Let p_i and $\mu_{ij} \ge 0$ be the shadow prices associated with constraints (8) – (9). The condition for social welfare maximization can be written,

$$W_u'(\overline{q}_i) = p_i \qquad \text{for } i = 1,...,n,$$

(10)

$$\sum_{j=1}^{n} [(p_i - \mu_{ij})Y_{ij} - (p_j - \mu_{ji})Y_{ji}] = 0, \qquad \text{for} \quad i = 1, \cdots, n-1$$

(11)

$$\mu_{ij}[Y_{ij}(\theta_i - \theta_j) - \overline{P}_{ij}] = 0. \qquad \text{for} \quad 1 \le i, j \le n$$

(12)

By assumption $W_i'(\overline{q}_i) \ge 0$. Therefore, the condition (10) implies that $p_i \ge 0$.

3. A Market Mechanism

In this section, we describe the fundamental elements for a market mechanism and demonstrate how it resolves the externality problem. Kirchoff's laws cause misalignment between the contract path of power transfer and the physical paths of power flow, leading to uncompensated use of parts of the network that lie outside of the contract path. As a result, the private cost and the social cost diverge from each other in electricity transactions. Traditionally, these externalities are handled through the coordination by a central system operator in a power pool formed by vertically integrated utilities. An alternative approach that we will describe below is to design new property rights and create markets for these rights so that the external effects associated with a transaction can be internalized in private decisions. We assume that the market operates within a power pool.

The market mechanism builds on tradable transmission capacity rights and a trading rule that governs the exchange of these rights within the power pool. To properly define transmission capacity rights, we unbundle the transmission network into directed links, each representing the capability to carry power along a power line in a specified direction. Power transfer could be thought of as a productive activity that takes as input the line capacity of some transmission links and produces as output the line capacity of some other links. The definition of transmission capacity rights is motivated by the idea of simulating the actual power flows on individual links as market transactions take place.

Definition: A transmission capacity right entitles its owner to the right to send a unit of power through a specific transmission line in a specific direction according to the trading rule. A fixed set of transmission capacity rights, $\overline{P} = \{\overline{P}_{ij} | 1 \leq i, j \leq n\}$, are issued for each link (i,j) and these rights are tradable.

The trading rule specifies the quantities of various transmission capacity rights that traders must acquire in order to complete an electricity transaction. Electricity trades are typically cast in terms of transferring power from one node to another node in the network without specifying the actual power flow paths. An essential role of the trading rule is to codify the relationship between power transfers from one point to another and power flows on individual links so that the transmission capacity rights can be meaningfully enforced. Without loss of generality, we may arbitrarily designate a node, say n, as a base point and only need to define the trading rule that governs the transactions between the base point and every other node in the network.[2]

Definition: The trading rule consists of a set of loading factors $B = \{\beta_{ij}^k | 1 \leq i, j, k, \leq n\}$ where the value β_{ij}^k represents the quantity of transmission capacity rights on link (i,j) that a trader needs to acquire in order to transfer a unit of power from node k to node n.

The above trading rule is sufficiently general to support bilateral transactions, because the terms of trade for bilateral transactions between an arbitrary pair of nodes could be defined by combining these coefficients. For instance, to transfer one unit of power from node k to node l requires $\beta_{ij}^{kl} \equiv \beta_{ij}^k - \beta_{ij}^l$ units of transmission capacity rights on link (i,j).

The market mechanism can thus be summarized as $(\overline{P}, B)$. An efficient design of these terms will be derived below but can be motivated through interpretations as follows: $\overline{P}_{ij}$ can be seen as the physical capacity limit of transmission line (i,j), and β_{ij}^k can be seen as the loading factor, or the increase in power flow, on link (i,j) due to the injection of a unit of power at node k.

An efficient trading rule

We now consider the design of an efficient trading rule which will induce a socially optimal allocation. First, the power flow equations (4) can be rewritten in the following matrix form,

$$
\begin{bmatrix} q_1 \\ q_2 \\ q_{n-1} \end{bmatrix} = \begin{bmatrix} \sum_{j=1}^{n} Y_{1j} & -Y_{12} & \cdots & -Y_{1(n-1)} \\ -Y_{21} & \sum_{j=1}^{n} Y_{2j} & \cdots & -Y_{2(n-1)} \\ \vdots & \vdots & \ddots & \vdots \\ -Y_{(n-1)1} & -Y_{(n-1)2} & \cdots & \sum_{j=1}^{n} Y_{(n-1)j} \end{bmatrix} \begin{bmatrix} \theta_1 \\ \theta_2 \\ \vdots \\ \theta_{n-1} \end{bmatrix} \tag{13}
$$

Note that there are only n-1 independent equations in (4), because with zero losses, $\sum_{i=1}^{n} q_i = 0$, and at the base point, $\theta_n \equiv 0$. Let γ denote the matrix on the right side of (13) and $A = [\alpha_{ij}]$ denote its inverse. Solving (13) yields,

$$
\theta_i = \sum_{j=1}^{n-1} \alpha_{ik} q_j, \qquad \text{for} \quad i = 1, \cdots, n-1 \tag{14}
$$

Therefore, we may write the loading factors as

$$
\beta_{ij}^k = \frac{\partial P_{ij}}{\partial q_k} = Y_{ij} \left(\frac{\partial \theta_i}{\partial q_k} - \frac{\partial \theta_j}{\partial q_k} \right)
$$

$$
= Y_{ij}(\alpha_{ik} - \alpha_{jk}) \qquad \text{for i, j, k} = 1, \ldots, n, \tag{15}
$$

where we define $\alpha_{in} = \alpha_{in} = 0$, for $i = 1, \ldots, n$.

Competitive equilibrium and social optimum

Let us denote by π_{ij} the price of the transmission capacity right for link (i,j) and the price of electricity at node i by p_i, for i, j=1,...,n. Let $q_{ij} \in R+$ denote the amount of power transferred from node i to node j. Then we have, $q_i = \sum_{j=1}^{n} (q_{ij} - q_{ji})$ for i = 1,...,n. Denoting by **p, q** and π vectors of (p_i), (q_i), and (π_{ij}) for $1 \leq i, j \leq n$.

Power transfer can be thought of as a joint production process which takes the generated power and transmission capacity rights as inputs and produces the delivered power as well as some other transmission capacity rights as outputs. As a consequence, each transaction consumes some transmission capacity rights and produces others at the same time. The transmission charge could be negative, when a transaction entails counter flows on some congested links and produces transmission capacity rights on those links that turn out to have a higher market value than those consumed. This insight can be restated more precisely in mathematics: A contract to transfer q_{kl} units of power from node k to node l requires a bundle of transmission capacity rights $\{\beta_{ij}^{kl} q_{kl} | 1 \leq i, j \leq n\}$. The transmission charge, which equals $\sum_{i=1}^{n} \sum_{j=1}^{n} \pi_{ij} \beta_{ij}^{kl} q_{kl}$, may be either positive or negative, depending on the magnitudes of β_{ij}^{kl} and the prices of transmission capacity rights, π_{ij}.

To simplify matters, we define the nodal injection charge as $v_k \equiv \sum_{i=1}^n \sum_{j=1}^n \pi_{ij} \beta_{ij}^k$, which is the sum of the economic rent associated with injecting a unit of power at node k, for k=1,...,n-1. Therefore, the transmission charge for power transfer from node k to node l equals $v_{kl} \equiv v_k - v_l$. As we will show below, $p_k + v_k = p_n$ in equilibrium for k=1,...,n-1. We now introduce the definition of competitive equilibrium.

Definition: A competitive equilibrium is a vector (p, q, π) that satisfies the following three conditions:

A. The market at each node is in equilibrium. That is, the price of electricity must be equal to the marginal cost and the marginal benefit at each node,

$$W_i'(\overline{q}_i) = p_i, \quad \text{for} \quad i = 1, \cdots, n, \tag{16}$$

B. There should be no positive profit to be made by transfer of power from one node to another, i.e.,[3]

$$p_l = p_k + \sum_{i=1}^n \sum_{j=1}^n \pi_{ij} \beta_{ij}^{kl}. \tag{17}$$

C. The price for transmission capacity rights is zero when there is excess supply:[4]

$$\pi_{ij} \left[\sum_{k=1}^n \sum_{l=1}^n \beta_{ij}^{kl} q_{kl} - \overline{P}_{ij} \right] = \pi_{ij} \left[\sum_{k=1}^n \beta_{ij}^k q_k - \overline{P}_{ij} \right] = 0. \quad \text{for} \quad 1 \le i, j \le n. \tag{18}$$

In Proposition 1, we demonstrate that the market mechanism can resolve the externalities associated with loop flows and restore market efficiency. This implies that once a properly designed system of transmission capacity rights and trading rule is in place, efficient transmission prices will result from competitive bidding. There is no need for a central agency to compute transmission prices in order to implement spot markets in an electric network.

Proposition 1. Under the market mechanism $(\overline{P}, B)$, a competitive equilibrium is socially optimal.

The proof is provided in the Appendix.

4. A Dynamic Trading Process

In this section, we investigate the dynamic behavior of the market mechanism . A dynamic trading process involves two types of activity: transmission bidding and electricity trading. First, given the electricity demands and supplies as well as electricity prices,[5] transmission capacity rights holders set the prices for transmission links through competitive auction. In equilibrium, the total economic rent will be maximized subject to the constraint that these prices will not render the prevailing transactions of electricity unprofitable. (The profitability constraint is necessary to ensure positive demands for these rights.) If the total economic rent is not maximized, there will be merchandising opportunities for traders to purchase a bundle of rights and then resell them for a profit. Let N be the set of all the transmission links in the network. We define the set of congested links as

$$C \equiv \{(i, j) \in N \mid \sum_{k=1}^{n-1} \beta_{ij}^k q_k = \overline{P}_{ij}\} \text{ and its complement, } \overline{C} \equiv N - C, \text{ represents}$$

the set of uncongested links. Then, the rent maximizing prices for transmission links, $\{\pi_{ij}\}$, as determined by market trading, would solve the following linear program:

$$R(p, q) \equiv \max_{\pi_{ij}} \sum_{i=1}^{n} \sum_{j=1}^{n} \pi_{ij} \overline{P}_{ij}, \tag{19}$$

subject to

$$p_n - p_k - \sum_{i=j}^{n} \sum_{j=1}^{n} \pi_{ij} \beta_{ij}^k \begin{cases} \geq 0, & \text{if } q_k \geq 0, \\ \leq 0, & \text{if } q_k \leq 0, \end{cases} \tag{20}$$

$$\text{for } k = 1, \cdots, n-1.$$

$$\pi_{ij} \begin{cases} \geq 0, & \text{if } (i, j) \in C, \\ = 0, & \text{if } (i, j) \in \overline{C}. \end{cases} \tag{21}$$

The condition in (20) requires that the transactions among electricity traders will not become unprofitable. The condition in (21) requires that the price of a transmission capacity right must be zero, if there is an excess supply of the right.

Propositions 2 and 3 below state that the merchandising surplus, which equals $-\sum_{k=1}^{n} p_k q_k$, is always greater than the total economic rent , and they are equal if and only if a competitive equilibrium is attained. Therefore, the trading rule is inherently "revenue adequate" in the sense that if all power transactions are handled by a central agency, the revenue would always be sufficient to cover the payments to the owners of transmission assets, and it would break even only at a competitive equilibrium.

Proposition 2. $R(p,q) \leq -\sum_{k=1}^{n} p_k q_k$; and $R(\mathbf{p},\mathbf{q}) = -\sum_{k=1}^{n} p_k q_k$ only when equality holds for every k in (20).

Proposition 3. $R(\mathbf{p},\mathbf{q}) = -\sum_{k=1}^{n} p_k q_k$, if and only if $(\mathbf{p},\mathbf{q},\pi)$ is a competitive equilibrium.

Next, in electricity trading, we assume that traders will exploit profitable arbitrage opportunities continuously. A set of marginal transactions are represented by a vector of differential net demands (**dq**) such that

$$\sum_{k=1}^{n} dq_k = 0. \tag{22}$$

A profitable arbitrage is defined as a set of marginal transactions that satisfy the following profitability and feasibility conditions:

$$-\sum_{k=1}^{n} p_k dq_k > 0, \tag{23}$$

and

$$\sum_{k=1}^{n} \beta_{ij}^{k} dq_k \leq 0 \qquad \text{for } (i, j) \in C. \tag{24}$$

The inequality (23) states that the transaction yields a positive net profit. The inequality (24) indicates that the transaction imposes non-positive demands for transmission capacity rights on congested transmission links.

Therefore, in this dynamic trading process, while the transmission capacity rights holders will maximize the total economic rent by exploiting merchandising opportunities through competitive auction, electricity traders will continuously exploit profitable arbitrage opportunities. A relevant question is whether this process will converge to an equilibrium. Proposition 4 suggests that the auction of transmission capacity rights provides useful information regarding the profitable arbitrage opportunities.

Proposition 4. $R(\mathbf{p},\mathbf{q}) < -\sum_{k=1}^{n} p_k q_k$, if and only if there exists a profitable arbitrage, $(dq_1,...,dq_n)$, that satisfies $(22) - (24)$.

Proposition 4 demonstrates that as long as the merchandising surplus exceeds the total economic rent, a profitable arbitrage can always be found. As we will show below, as a result of such an arbitrage, the social welfare increases. Therefore, these arbitrage opportunities resemble an invisible hand that guides the traders in pursuing their own gains to contribute to the social welfare. The trading process would continue until the total economic rent equals the merchandising surplus, i.e.,

$R(\mathbf{p}, \mathbf{q}) = -\sum_{k=1}^{n} p_k q_k,$. According to Proposition 3, this implies that the process converges to a competitive equilibrium. Therefore, the above dynamic trading process is Lyapunov stable in the sense that the social welfare, which serves as a

Lyapunov function, increases monotonically as the trading evolves toward a competitive equilibrium.

Proposition 5. The dynamic trading process is Lyapunov stable and converges to a competitive equilibrium.

In the above dynamic trading process, the market equilibrium emerges as a consequence of spontaneous interactions of electricity trading and transmission bidding. In equilibrium, we have $p_k + v_k = p_n$, for $k=1,...,n-1$.

5. An Illustrative Example

In this section, we present a numerical example that simulates how a dynamic trading process would converge to a competitive equilibrium. For illustrative purposes, we consider a simple electric power network with three nodes as shown in Figure 1, in which nodes 1 and 2 are supply nodes, and node 3 is a demand node. Table 1 shows the specific assumption about the marginal cost and inverse demand functions for off-peak, normal, and peak periods. We asssume that the rated capacities for lines (1,2), (1,3) and (2,3) are 100 MW, 300 MW and 220 MW, respectively. We further assume that the impedances of these lines are identical, i.e. $Y_{ij} \equiv Y$, for $i,j=1,2,3$. Under the above assumptions, we can obtain the trading rule as shown in Table 2. For example, in order to transfer 1 MW from node 1 to node 3, the trader will have to acquire 1/3 MW of the transmission capacity right for link (1,2), 1/3 MW for link (2,3), and 2/3 MW for link (1,3); at the same time, the transaction will generate for the trader a credit of 1/3 MW of transmission capacity right for each of the two links (2,1) and (3,2) plus 2/3 MW of the right for link (3,1).

In the following, we simulate a dynamic trading process in discrete steps, each of which is illustrated with a Figure as indicated in the parentheses.

Step 0 (Figure 2)

To begin, we assume that the system is in competitive equilibrium during an off-peak period. During off-peak, the network is not congested. At competitive equilibrium, a uniform market price of $22.86/MWh prevails through out the network. The transmission congestion rent is zero. The power flows are determined by Kirchoff's laws. For example, given that $q_1 = 257$ MW and $q_2 = 29$ MW, Kirchoff's laws dictate that two thirds of the power supplied at node 1 and one third of the power supplied at node 2 will flow through link (1,3). Therefore, the power flow on link (1,3) is 181 MW ($= 257 \times (2/3) + 29 \times (1/3)$).

Step 1 (Figure 3)

Next, suppose that the demand and supply functions are suddenly changed to those for a normal period. With output unchanged, the immediate effect of such a change is that while the price at node 1 stays at $22.86/MWh, the prices at nodes 2 and 3 rise to $p_2 = \$37.86/MWh$ and $p_3 = \$82.86/MWh$. With zero transmission charges, there obviously are profitable trading opportunities. One of the most profitable possibilities is to buy electricity from node 1 and sell it to node 3.

Step 2 (Figure 4)

The above trade could continue until $q_1 = 329$ MW and $\bar{q}_3 = 358$ MW, when link (1,2) becomes congested. The prices of electricity at nodes 1 and 3 would be changed accordingly to $p_1 = \$26.45/\text{MWh}$ and $p_3 = \$68.40 /\text{MWh}$. Then, through competitive auction, the price of the transmission capacity right for link (1,2) could be bid up to $\$126/\text{MWh}$ to extract the maximum transmission congestion rent. The rent maximization problem can be stated as follows:

$$\max_{\pi_{12}} 100\pi_{12}$$

subject to

$$26.45 + \pi_{12}/3 \leq 68.40,$$

and

$$37.86 - \pi_{12}/3 \leq 68.40.$$

Since the total congestion rent (= $2600/h) is less than the merchandising surplus (= $14687 /h), there are still arbitrage opportunities that remain to be exploited. One such arbitrage opportunity is to transfer an identical amount of power simultaneously from nodes 1 and 2 to node 3. According to the trading rule, the transfer of 1 MW from node 2 to node 3 creates a 1/3 MW transmission capacity right credit for link (1,2), which exactly offsets the transmission capacity right needed for transferring power from node 1 to node 2. Therefore, the arbitrage imposes no net requirement for the transmission capacity right for the congested link. Such an arbitrage would initially yield a positive profit of $36.25/MWh (=68.40 - (26.45+37.86)/2).

Step 3 (Figure 5)

As the arbitrage proceeds, the power flows would increase, but the price of the transmission capacity right for link (1,2) as well as the congestion rent for link (1,2) would fall. When the net supplies at nodes 1 and 2 reach 400 MW and 100 MW, respectively, the price of the transmission capacity right for link (1,2) becomes $30 /MWh. The profit margin drops to $2.5/MWh (=40 - (30+45)/2), but is still positive. However, the power flow on link (1,3) reaches the capacity limit, which prevents further arbitrage. In the end, we have $p_1 = 30$, $p_2 = 45$ and $p_3 = 40$.

Step 4 (Figure 6)

As link (1,3) is loaded up to capacity limit, it could command a positive congestion rent. The prices of transmission capacity rights for links (1,2) and (1,3) are determined by competitive auction, which yields maximum congestion rent as follows:

$$\max_{\pi_{12},\pi_{13}} 100\pi_{12} + 300\pi_{13}$$

subject to

$$p_1 + \pi_{12}/3 + 2\pi_{13}/3 \le p_3,$$

and

$$p_2 - \pi_{12}/3 + \pi_{13}/3 \le p_3.$$

With the prevailing prices, we obtain $\pi_{12} = \$20$ /MWh and $\pi_{13} = \$5$ /MWh for a total transmission congestion rent of \$3500/h. At this point the merchandising surplus equals the transmission congestion rent. Therefore, a competitive equilibrium is attained.

Note that although $p_2 > p_3$ at equilibrium, power is actually delivered from node 2 to node 3. According to the traditional interpretation, the transmission charge from node 2 to node 3 must be -\$5/MWh, a phenomenon which has been commonly viewed as one of the peculiarities of electric power network. There is nothing mysterious, however, once we interpret the results using the framework of a market for transmission capacity rights.

Step 5 (Figure 7)

Finally, we consider the case in which the demand and supply functions are changed to those for a peak period. If the quantities of demand and supply are fixed in the short term, the immediate market response would be changes in prices to $p_1 = \$30$ /MWh, $p_2 = \$40$/MWh, and $p_3 = \$65$/MWh. With these prices, we can solve the rent maximization problem in Step 4 and obtain $\pi_{12} = 0$ and $\pi_{13} = 52.5$. Once again, we have a situation in which the mechandising surplus (\$16500/h) is greater than the total congestion rent (\$15750/h), and thus new arbitrage opportunities are created. At first sight, it seems profitable to transfer power from node 2 to node 3. However, the congestion of the transmission link (1,3) renders this transaction infeasible. Instead, for each MW bought from node 2, one could sell 1/2 MW to node 3 and the remaining 1/2 MW to node 1 (i.e., to reduce the electricity output at node 1) for a profit of \$7.5/MWh.[6] These transactions do not require the transmission capacity right for link (1,3) and can be continued until link (2,3) becomes congested.

Step 6 (Figure 8)

At the end of the arbitrage, the nodal prices would become $p_1 = \$29$/MWh, $p_2 = \$44$/MWh and $p_3 = \$61$/MWh. As link (2,3) becomes congested, the rent maximization problem is restated as follows:

$$\max_{\pi_{13}, \pi_{23}} 300\pi_{13} + 220\pi_{23}$$

subject to

$$p_1 + 2\pi_{13}/3 + \pi_{23}/3 \le p_3,$$

and

$$p_2 + \pi_{13}/3 + 2\pi_{23}/3 \le p_3.$$

The equilibrium prices of transmission capacity rights are $\pi_{13} = \$47$/MWh and $\pi_{23} = \$2$/MWh. The total transmission congestion rent is \$14540, which equals the merchandising surplus. Therefore, a competitive equilibrium is attained.

The results of the dynamic trading process are summarized in Table 3. Note that the equilibrium price at node 1 during the peak period is lower than during the normal period, and the congestion rent rises sharply during the peak period. This is because as link (1,3) becomes congested, the demand at node 3 can only be met by increasing the output at node 2 and, at the same time, decreasing the output at node 1. Therefore, although $p_1 < p_2$, link (1,2) remains uncongested.

Alternative implementation plans

Thus far, we have focused on the basic principles of relying on a market mechanism to resolve market failures due to loop flow externalities. Given the complexity of the industry, any efficient implementation strategy must take into account the historical path, existing market structure, and technical uncertainties. A detailed investigation of these and other issues related to implementation is beyond the scope of this paper. See Chao and Peck (1997). In the following, we provide some preliminary thoughts about how the theory could be brought to bear on practice in relation to transmission pricing.

We have demonstrated that a set of appropriately defined transmission capacity rights and a trading rule would be sufficient to ensure an efficient decentralized market for electricity. Theoretically, these principles could provide a useful benchmark for evaluation of other more practical approaches to transmission pricing and access, such as zonal pricing and point-to-point service.[7] In practice, depending on how the prices of electricity and the transmission charges are actually determined, three alternative implementation plans could be envisaged:

A. The electricity prices and the transmission charges are determined jointly by a central agency based on the social welfare maximization model in which the benefit and cost functions are constructed from bids received from electricity buyers and sellers.[8]

B. The nodal injection charges (v_k) are set by a central agency based on the rent maximization model in (19) - (21), while the electricity prices (p_k) are determined by a dynamic trading process.

C. The injection charges and the electricity prices are both determined in a decentralized dynamic trading process.

Each of these plans represents a fundamental departure from the current practice in their use of market to organize the spot trading of electricity. While these plans embody the same principles, they vary with an increasing degree of decentralization as we move from plan A to plan C, implying rather different market organizations for the spot trading of electricity. While plan C is conceptually most appealing, plan A may be a more practical first step, for it entails relatively less dramatic departure from the existing institutional structure.

6. Conclusions

The externalities caused by loop flows in a transmission network represent a critical barrier to the introduction of competition into the electric power industry. In this paper, we investigate the design of a market mechanism for electric power transmission that consists of tradable transmission capacity rights and a trading rule. We demonstrate the basic principles of using a market mechanism for the resolution of externalities and the efficiency properties of a dynamic trading process for electric power transmission. Conceptually, these principles represent a fundamental departure from those underlying the current practice, raising the prospect of a competitive market for transmission services and a competitive spot market for electricity. In practice, concepts from the new approach could be applied to support the implementation of alternative proposals for restructuring a power pool and may serve as a useful benchmark for other approaches to market designs.

In the long term, the expansion of a transmission network is characterized by potential economies of scale and externalities and requires alternative mechanisms. How the long term investment incentives should be structured and how the investment decisions will interact with the market expectations in the short run is a fundamental question that remains to be answered in future research.

Appendix

Proposition 1. Under the market mechanism $(\overline{P}, B)$, a competitive equilibrium is socially optimal.

Proof of Proposition 1: We first note that (8) is ensured by Kirchoff's laws, and (10) follows directly from (16). Using (14) and (15), we may derive the demand for transmission capacity rights as follows,

$$
\begin{aligned}
\sum_{k-1}^{n} \beta_{ij}^{k} q_{k} &= \sum_{k=1}^{n} Y_{ij} (\alpha_{ik} - \alpha_{jk}) q_{k} \\
&= Y_{ij} (\theta_{i} - \theta_{j}).
\end{aligned}
\tag{25}
$$

Since the quantity of tradable transmission capacity rights demanded can not exceed the total quantity issued, (9) holds.

We may interprete the equilibrium price as the shadow price of the transmission constraint, i.e., $\mu_{ij} = \pi_{ij}$. It follows from (18) and (25) that (12) holds.

Next, let us consider the transaction of buying one unit of power at node n, the base point, and selling it at node 1. Note that by definition, we have, $\beta_{ij}^{nl} = -\beta_{ij}^{l}$. The equilibrium condition in (17) can be written,

$$p_1 = p_n + \sum_{i=1}^{n} \sum_{j=1}^{n} \pi_{ij} \beta_{ij}^{nl}$$

$$= p_n - \sum_{i=1}^{n} \sum_{j=1}^{n} \mu_{ij} \beta_{ij}^{l}$$

$$= p_n - \sum_{i=1}^{n} \sum_{j=1}^{n} \mu_{ij} Y_{ij} (\alpha_{il} - \alpha_{jl})$$

$$= p_n + \sum_{i=1}^{n} \sum_{j=1}^{n} \alpha_{il} (\mu_{ji} Y_{ji} - \mu_{ij} Y_{ij}), \quad \text{for } l = 1, \cdots, n-1.$$

(26)

Equations (26) can be rewritten in a matrix form as follows,

$$
\begin{bmatrix} p_1 \\ p_2 \\ \vdots \\ p_{n-1} \end{bmatrix} = \begin{bmatrix} 1 \\ 1 \\ \vdots \\ 1 \end{bmatrix} p_n + \begin{bmatrix} \alpha_{11} & \alpha_{21} & \cdots & \alpha_{(n-1)1} \\ \alpha_{12} & \alpha_{22} & \cdots & \alpha_{(n-1)2} \\ \vdots & \vdots & \ddots & \vdots \\ \alpha_{1(n-1)} & \alpha_{2(n-1)} & \cdots & \alpha_{(n-1)(n-1)} \end{bmatrix} \times
$$

(27)

$$
\begin{bmatrix} \sum_{j=1}^{n} (\mu_{j1} Y_{j1} - \mu_{1j} Y_{1j}) \\ \sum_{j=1}^{n} (\mu_{j2} Y_{j2} - \mu_{2j} Y_{2j}) \\ \vdots \\ \sum_{j=1}^{n} (\mu_{j(n-1)} Y_{j(n-1)} - \mu_{(n-1)j} Y_{(n-1)j}) \end{bmatrix}
$$

Multiplying both sides of (27) by Y_T, and using the fact that $A = Y_{-1}$, we obtain (11). We have thus proven that the competitive equilibrium is socially optimal. **QED**

Proposition 2. $R(\mathbf{p}, \mathbf{q}) \leq -\sum_{k=1}^{n} p_k q_k$; and $R(\mathbf{p}, \mathbf{q}) = -\sum_{k=1}^{n} p_k q_k$; only when equality holds for every k in (20).

Proof of Proposition 2: Multiplying (20) by q_k and summing it over k, we obtain,

$$\sum_{k=1}^{n} \sum_{i=1}^{n} \sum_{j=1}^{n} \pi_{ij} \beta_{ij}^{k} q_k = \sum_{i=1}^{n} \sum_{j=1}^{n} \pi_{ij} \overline{P}_{ij}$$

$$\leq -\sum_{k=1}^{n} (p_k - p_n) q_k$$

(28)

$$= -\sum_{k=1}^{n} p_k q_k,$$

where π_{ij}, for $1 \le i,j \le n$ satisfy (21). Note that in (28), we incorporate the observation that the inequality in (20) holds by default when k=n. The right hand sides and left hand sides in (28) are equal only when equality holds in (20) for every k. Choosing (π_{ij}) that maximizes the left hand side of (28), we have $R(\mathbf{p}, \mathbf{q}) \le -\sum_{k=1}^{n} p_k q_k$. **QED**

Proposition 3. $R(\mathbf{p}, \mathbf{q}) = -\sum_{k=1}^{n} p_k q_k$, if and only if $(\mathbf{p}, \mathbf{q}, \pi)$ is a competitive equilibrium.

Proof of Proposition 3: If $(\mathbf{p}, \mathbf{q}, \pi)$, is a competitive equilibrium, we have, by setting $l = n$ in (17) and rearranging terms,

$$\sum_{i=1}^{n} \sum_{j=1}^{n} \pi_{ij} \beta_{ij}^{k} = p_n - p_k \qquad \text{for } k = 1, \cdots, n-1, \tag{29}$$

where (π_{ij}) satisfies (21). Multiplying both side of (29) by q_k and summing over k, we obtain,

$$\sum_{k=1}^{n} \sum_{i=1}^{n} \sum_{j=1}^{n} \pi_{ij} \beta_{ij}^{k} q_k = \sum_{k=1}^{n} (p_n - p_k) q_k$$

$$= -\sum_{k=1}^{n} p_k q_k. \tag{30}$$

Since (30) implies $R(\mathbf{p}, \mathbf{q}\}) \ge -\sum_{k=1}^{n} p_k q_k$, we obtain from Proposition 2 $R(\mathbf{p}, \mathbf{q}) = -\sum_{k=1}^{n} p_k q_k$.

Conversely, suppose that $R(\mathbf{p}, \mathbf{q}) = -\sum_{k=1}^{n} p_k q_k$. It follows from Proposition 2 that equality must hold for every k in (20). This proves that $(\mathbf{p}, \mathbf{q}, \pi)$ is a competitive equilibrium. **QED**

Proposition 4. $R(\mathbf{p}, \mathbf{q}) < -\sum_{k=1}^{n} p_k q_k$, if and only if there exists a profitable arbitrage, $(dq_1,...,dq_n)$.

Proof of Proposition 4: The proof is a direct application of the Separating Hyperplane Theorem[9] Let us first consider the following simultaneous equations in π_{ij} for $(i,j) \in C$:

$$\sum_{i=1}^{n} \sum_{j=1}^{n} \pi_{ij} \beta_{ij}^{k} = p_n - p_k \qquad \text{for } k = 1, \cdots, n-1. \tag{31}$$

By Proposition 3, if $R(\mathbf{p}, \mathbf{q}) < -\sum_{k=1}^{n} p_k q_k$, the simultaneous equations in (31) can not have a nonnegative solution $(\pi_{ij}) \geq 0$. Let us define $\gamma^k \equiv p_n - p_k$. This means that the $(n-1)$-vector $(\gamma^1, \ldots, \gamma^{n-1})$ lies outside the convex cone formed by the vectors $(\beta_{ij}^1, \cdots, \beta_{ij}^{n-1})$ for $(i,j) \backslash \in$ C. Then, from the Separating Hyperplane Theorem, this will be the case, if and only if there exists a vector $(dq_1, \ldots, dq^{n-1})$ which satisfies the following inequalities:

$$\sum_{k=1}^{n-1} \beta_{ij}^k dq_k \leq 0 \quad \text{for } (i, j) \in C.$$

(32)

and

$$\sum_{k=1}^{n-1} (p_n - p_k) dq_k > 0.$$

(33)

Note that (32) and (24) are equivalent, because $\beta_{ij}^n \equiv 0$.
The expression (33) can be rewritten as follows,

$$0 < \sum_{k=1}^{n-1} (p_n - p_k) dq_k$$

$$= p_n \sum_{k=1}^{n-1} dq_k - \sum_{k=1}^{n-1} p_k dq_k$$

(34)

$$= -\sum_{k=1}^{n} p_k dq_k.$$

Therefore, (23) holds. **QED**

Proposition 5. The dynamic trading process is Lyapunov stable and converges to a competitive equilibrium.

Proof of Proposition 5: From Proposition 4, as long as $R(\mathbf{p}, \mathbf{q}) < -\sum_{k=1}^{n} p_k q_k$, during the trading process, there always exists a profitable trading opportunity $(dq_1, \ldots, dq_n)$. As a result of such an arbitrage, the social welfare is changed by

$$dW = \sum_{k=1}^{n} W_k'(\overline{q}_k) d\overline{q}_k$$

$$= \sum_{k=1}^{n} p_k d\overline{q}_k$$

(35)

$$= -\sum_{k=1}^{n} p_k dq_k > 0.$$

Therefore, the social welfare increases monotonically during the dynamic trading process until the arbitrage opportunities are exhausted. If the trading process converges to $(\mathbf{p^*}, \mathbf{q^*}, \pi^*)$, then from Propositions 4 and 6, we obtain $R(\mathbf{p^*}, \mathbf{q^*}) = -\sum_{k=1}^{n} p_k^* q_k^*$. By Proposition 3, this implies that $(\mathbf{p^*}, \mathbf{q^*}, \pi^*)$, is a competitive equilibrium. **QED**

References

Chao, Hung-po and Stephen Peck (1996) ``A Market Mechanism for Electric Power Transmission", *Journal of Regulatory Economics*, July, Vol. 10, No. 1.

Chao, Hung-po and Stephen Peck (1997) "An Institutional Design for an Electricity Contract Market with Central Dispatch", *Energy Journal*, January.

Coase, R. (1960), "The Problem of Social Cost", *The Journal of Law and Economics* 3, October, pp. 1-44.

Elgerd, O.I. (1982) *Electric Energy Systems and Theory*, 2nd ed. New York: McGraw-Hill Book Company.

Gale, David (1960), "The Theory of Linear Economic Models," The University of Chicago Press, Chicago, Illinois.

Graves, Frank (1995), "A Primer on the Economics of Electric Power Flow", EPRI TR-104604, Electric Power Research Institute, Palo Alto, California

Hogan, William (1992), "Contract Networks for Electric Power Transmission," Journal of Regulatory Economics, Vol. 4, pp. 211-242, September.

Hogan, William (1993), "Electric Transmission: A New Model for Old Principles", The Electricity Journal}, March

Joskow, Paul and Richard Schmalensee (1983), *Markets for Power*, The MIT Press, Cambridge, Massachusetts.

Schweppe, Fred C. (1980), "Homeostatic Utility Control", Technical Paper, MIT Energy Lab.

Schweppe, F.C., M.C. Caramanis, R.D. Tabors and R.E. Bohn (1988) *Spot Pricing of Electricity*, Kluwer Academic Publishers: Norwell, MA.

Wu, F., P. Varaiya, P. Spiller and S. Oren (1994), "Folk Theorems on Transmission Access: Proofs and Counter Examples", Program on Workable Energy Regulation (POWER), University of California.

Endnotes

[1] Electric Power Research Institute. The authors are indebted to Charles Clark, William Hogan, Shmuel Oren, Pravin Varaiya, Robert Wilson, Felix Wu, and two referees for

helpful comments and suggestions. This paper does not represent the views of EPRI or its members. The authors remain solely responsible for the errors in this paper.

[2] The designation of base point is largely a mathematical convenience. However, in practice, it may preferably be located at a large load bus near the center of the network where large volumes of power transactions take place so that the market price of electricity is readily available.

[3.] The cost of a transaction includes the price of power purchased, p_k, and the sum of the prices of transmission capacity rights over all the links used, $\sum_{i=1}^{n} \sum_{j=1}^{n} \pi_{ij} \beta_{ij}^{kl}$. Strictly speaking, the no positive profit condition should be stated as an inequality:

$$p_l \leq p_k + \sum_{i=j}^{n} \sum_{j=1}^{n} \pi_{ij} \beta_{ij}^{kl}.$$

However, since $\beta_{ij}^{kl} = -\beta_{ij}^{lk}$, we may switch the indices k and l in the above inequality and show that the equality in (17) holds.

[4] The total demand for the transmission capacity right for link (i,j) is obtained by summing the individual demand $\beta_{ij}^{kl} q_{kl}$ over all power transactions:

$$\sum_{k=1}^{n} \sum_{l=1}^{n} \beta_{ij}^{kl} q_{kl}.$$

[5] We assume that initially, a uniform price prevails at each node, such that $W_k'(\bar{q}_k) = p_k$, for k=1,...,n.

[6] These transactions effect an increase in the expensive supply from node 2 to replace the less expensive supply at node 1. This phenomenon is known as out-of-merit dispatch.

[7] Zonal pricing is a simplified scheme that approximates the nodal injection charges (v_k) within a specified zone by a uniform price. Point-to-point service is an alternative scheme that approximates the spot transmission charge v_{kl} by a posted rate.

[8] Since the prices are computed at equilibrium, the transmission charge for power transfer from node k to node l ($v_{kl} = v_k - v_l$) equals the difference of the corresponding nodal electricity prices ($p_l - p_k$). Therefore, this implementation plan is similar to Hogan's (1992) with the exception that the economic rents are allocated on the basis of transmission capacity rights rather than transmission congestion contracts.

[9] A version of the Separating Hyperplane Theorem can be stated as: Suppose that A is an m x n matrix, and b is an m-vector. Then the simultaneous equation Ax=b has no nonnegative solution with $x \in R_+^n$, if and only if there exists a vector $y \in R^m$ with $y^T A \leq 0$ and $y^T b > 0$. (Gale 1960, p. 44.).

Chapter 5

THE BERLIN MANDATE: THE DESIGN OF COST-EFFECTIVE MITIGATION STRATEGIES[1]

Richard Richels, Electric Power Research Institute,
Jae Edmonds, Pacific Northwest Laboratory
Howard Gruenspecht, U.S. Department of Energy
and Tom Wigley, University Corporation for
Atmospheric Research

Abstract: *The Berlin Mandate calls for strengthening developed country commitments for limiting greenhouse gas emissions. This paper addresses a key issue in the current analysis and assessment phase -- the costs of proposals to limit CO2 emissions. Employing four widely-used energy-economy models, we explore the direct and indirect effects of alternative proposals on the global economy. We also examine the implications for atmospheric CO2 concentrations. We begin by examining an AOSIS-like proposal in which OECD countries agree to reduce CO2 emissions by 20% below 1990 levels by a specified date. We find that implementing such a proposal could be quite costly. Not surprisingly, OECD countries would be hardest hit. Their costs could be as high as several percent of GDP. The analysis also shows that because of trade effects, non-OECD countries would likely incur costs even when reductions are confined to the OECD. An economic slowdown in the OECD would affect the full range of developing country exports, and hence their economic growth. This would likely be the case for both oil-importing and oil-exporting developing countries. We then explore alternatives that are apt to be quite similar in terms of environmental benefits, but allow for flexibility in where and when emission reductions are made. We find that costs could be substantially reduced through international cooperation and the optimal timing of emission reductions. Indeed, such flexibility can reduce costs by more than 80%, potentially saving the international community trillions of dollars in mitigation costs. We find that reliance on more flexible alternatives reduces costs more effectively than adopting weaker, but still inflexible, commitments.*

John Weyant (ed.), ENERGY AND ENVIRONMENTAL POLICY MODELING. Copyright © 1998. Kluwer Academic Publishers. ISBN 0-7923-8348-6. All rights reserved.

1. Introduction

The Berlin Mandate calls upon the Parties to the United Nations Framework Convention on Climate Change to strengthen developed country commitments to reduce greenhouse gas emissions.[2] A number of proposals have been put forward. These range from slowing the current growth in emissions to sharp reductions below present levels. The choice is a difficult one. Acting too slowly risks irreversible environmental damages. Acting too aggressively risks imposing large, and perhaps unnecessary costs on the global economy. As noted by the Intergovernmental Panel on Climate Change (IPCC), the challenge is to develop a prudent hedging strategy in the face of climate-related uncertainties.[3]

The Framework Convention is the mechanism established by the international community for implementing precautionary measures. It recognizes that a sensible hedging strategy should be flexible, with ample opportunities for learning and mid-course corrections. Periodic reviews are required "in light of the best available scientific information on climate change and its impacts, as well as relevant technical, social and economic information." Based on these reviews, appropriate measures would be taken, including the adoption of new commitments.

Upon entering into force in 1994, the Convention established an initial (but non-binding) aim for developed countries to return emissions to their 1990 levels by 2000. At the first meeting of the Conference of the Parties (COP-1) in Berlin in April of 1995, it was determined that existing commitments under the Convention were inadequate. Further commitments for developed countries are to be negotiated, and prepared for approval at COP-3 in 1997.

While calling for new commitments, the Berlin Mandate does not specify what these commitments should be. Rather it seeks further analysis and assessment to guide and inform the decision making process. This paper addresses a key issue in the analysis and assessment phase -- the costs of proposals to limit CO2 emissions. Rather than rely on a single model, the analysis is based on independent runs of four widely-used energy-economy models.[4] In each instance, we explore both the direct and indirect effects on the global economy.

We also examine the impact of alternative proposals on atmospheric concentrations. The ultimate objective of the Framework Convention is "the stabilization of greenhouse gas concentrations in the atmosphere at a level that would prevent dangerous anthropogenic interference with the climate system."[5] Although the issue of what constitutes an appropriate limit has yet to be resolved, it is instructive to explore the implications that alternative emission pathways have for future concentrations.

We pay particular attention to the design of cost-effective mitigation strategies. The Framework Convention states that "policies and measures to deal with climate change should be cost-effective so as to ensure global benefits at the lowest possible costs." Adopting least-cost mitigation strategies will free up valuable resources for further addressing the threat of climate change or for meeting other societal needs. We explore two ways of promoting this objective. In the first, emission reductions take place *where* it is cheapest to do so. In the second, they take place *when* it is cheapest to do so.

A number of studies have suggested that the cost of emission reduction can be substantially reduced through international cooperation.[6] From a global perspective, it would be economically wasteful to incur high marginal abatement costs in one country when low-cost alternatives are available elsewhere in the world. We discuss ways to ensure that emission reductions are made where it is cheapest to do so, and explore the potential gains.

The timing of emission reductions can also influence costs. What is important in meeting a concentration target is cumulative -- not year-by-year -- emissions.[7] A particular concentration target can be met through a variety of emission time paths. Several studies have suggested that emission time paths that provide flexibility in making the transition away from fossil fuels will be less costly.[8] We examine the implications for the design of cost-effective mitigation strategies under the Berlin Mandate.

Mitigation costs are, of course, only part of the story. The more difficult question is the appropriate level of emissions abatement. This requires consideration of both costs *and* benefits. The present analysis is confined to the cost-side of the ledger. That is, we focus on the costs of emissions reduction. Policy makers will also want to know what they are buying in terms of reducing the undesirable consequences of global climate change. Such an analysis is beyond the scope of the present effort.

2. The models

The analysis employs four energy-economy models: CETA[9], EPPA[10], MERGE[11], MiniCAM[12]. These models reflect the recent trend towards hybrid modeling tools which incorporate features from both bottom-up and top-down approaches to energy modeling. On the supply-side, each model employs a bottom-up representation of the energy system. Energy technologies are described in process model detail (e.g., availability dates, heat rates, carbon emission coefficients, etc.). The technology vector includes both existing sources and new options that are likely to become available. Cost and performance constraints are adjusted for regional differences. A more top-down perspective is taken towards the balance of the economy. This is done using macroeconomic production functions that provide for substitution between capital, labor, and energy inputs.

The models provide a consistent way to examine alternative strategies for limiting CO2 emissions and to examine the impacts of higher energy prices on economic output. They can be used, for example, to estimate the increase in fossil fuel prices required to induce consumers to reduce emissions. They also can be used to analyze the possibility of significant regional differences in marginal abatement costs that would lead to opportunities for cost savings through international cooperation.

The models employ a general equilibrium formulation of the global energy and economic system. This allows us to examine the impacts of actions taken in one region on the economies of another. This is particularly important in the case of the Berlin Mandate. Constraints imposed on developed countries may have unexpected consequences for developing countries. For example, the international price of oil will be affected by the imposition of carbon constraints on oil importing countries.

While general equilibrium models are useful in tracing the long-term implications of a carbon constraint, they may ignore important short-term effects. This is because they assume full employment of the economy and instantaneous adjustment to policy shocks. The lack of attention to adjustment costs, means that these models may *understate* the short-run cost of economic shocks, particularly if these are large and unexpected.

On the other hand, some have argued that the exogenous specification of technology change tends to *overstate* the cost of a carbon constraint. This is an important issue in the energy policy debate -- one that is deserving of considerably more attention than it has received to date. It should be noted, however, that the direction of any bias is still unclear. An acceleration of energy related technical progress may be accompanied by a slowdown in labor and capital productivity improvements throughout the economy. To receive proper consideration, the issue of endogenous change must be examined on an economy-wide basis.[13]

Although similar in many respects, the models differ in important ways. For example, EPPA is a recursive rather than an intertemporal optimization model. EPPA and MERGE employ a "putty-clay" rather than a "putty-putty" approach to the vintaging of capital stocks (i.e., they explicitly recognize that one type of capital cannot be "transformed" into another, once it is put into place). And all models differ in regional disaggragation: CETA (2 regions), EPPA (12 regions), MERGE (5 regions) and MiniCAM (9 regions).[14]

The models also differ with respect to key inputs, e.g., population, per capita productivity trends, the fossil-fuel resource base, the cost and availability of long-term supply options, etc. Rather than try to impose a common set of driving assumptions, the choice of inputs was left to the discretion of the modeling teams. It was felt that, with a diverse set of energy futures, we would be better able to assess the robustness of our results.

3. Future emissions

We begin with an examination of how fossil fuel emissions are projected to grow in the absence of policy intervention. The costs of a carbon constraint are quite sensitive to the emissions baseline. The baseline describes how emissions will grow under existing policies. The higher the emissions baseline, the more carbon must be removed from the energy system to meet a particular target, and the higher become the costs.

Figure 1 compares baseline projections for our four models.[15] Note that in each instance emissions are projected to grow in the absence of policy intervention. This is the case for the OECD and for the world as a whole. This is consistent with the overwhelming majority of analyses recently reviewed by the IPCC.[16] Of the dozens of studies surveyed, all but a few showed a rising emissions baseline.

Figure 1. Carbon Emissions under Business As Usual

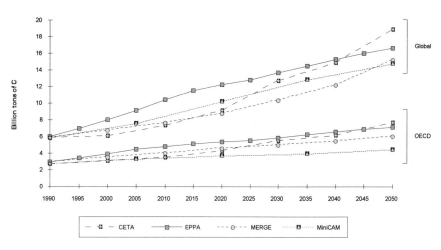

In reviewing these emission projections, several points are worth noting. First, although the annual growth rates are substantial -- between 1.5 percent and 2 percent -- they represent a marked slowing in the historical trend. Indeed, global emissions grew at an annual rate of approximately 3.5 percent between 1950 and 1990. In part, the slowdown is due to a projected decline in global economic growth. Since 1950, gross world product grew at an average annual rate of 2.9 percent. The projected growth rate for the next half century or so is closer to 2.5 percent. Also at work is the gradual decoupling between energy and GDP growth and a decoupling between CO2 emissions and energy use.

The differences in emission baselines should come as no surprise given the uncertainty over the period studied and thus the freedom the modelers had in the choice of input assumptions. Although it would be impractical to sort out all of the reasons for the differences, several factors have been identified as being particularly important when modeling future emissions.[17] High up on the list is economic growth. Those models with higher GDP growth rates tend to project higher emissions. Conversely, the more optimistic one is about the prospects for reducing energy intensity or the availability of low-cost carbon-free substitutes, the lower the CO2 growth rate.

Although the models differ on the cost and availability of supply and demand side alternatives, it should be noted that each includes some "no regrets" emission reduction options. These are alternatives that would be worth adopting apart from climate considerations. A growing emissions baseline does not imply the absence of economically competitive alternatives to fossil fuels. It only means that such options are in insufficient supply to arrest the growth in carbon emissions.

The focus of the Berlin mandate is on developed country emissions. Negotiators will be interested in how a particular proposal changes the emissions baseline. We start by examining a case in which OECD countries return emissions to 1990 levels by the year 2000, reduce them by an additional 20% by 2010, and hold them at that level thereafter. This is similar in many respects to the proposal put forward by the Alliance of Small Island States (AOSIS).[18] For the present analysis, we place no constraints on non-OECD emissions.

Figure 2 shows the implications for global emissions. An AOSIS-like proposal may slow the growth in global emissions, but it is unlikely to stabilize them at anywhere near present levels. This is because non-OECD countries currently account for over half of the global total and their share is expected to grow. The implications for climate policy are clear. Stabilization of global emissions will eventually require the participation of developing countries.

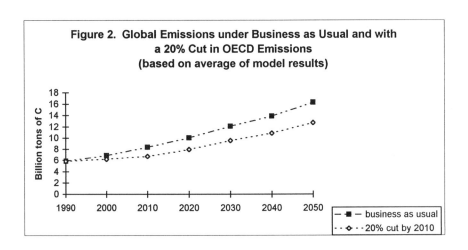

Figure 2. Global Emissions under Business as Usual and with a 20% Cut in OECD Emissions (based on average of model results)

4. The costs of alternative commitments

We next turn to the issue of costs. In recent years, a number of studies have highlighted the potential role of international cooperation and flexible timing in reducing the costs of a carbon constraint.[19] To explore the implications for the Berlin Mandate, we first estimate the costs of adopting the AOSIS-like proposal described above. We then examine three variants. Each results in the same cumulative emissions, but there are significant differences in the geographical location and timing of the emission reductions.

Before proceeding, one caveat is in order. We use trade in emission rights to examine the potential gains from international cooperation. By allowing such trade, we ensure that, at a given point in time, emission reductions are made where it is cheapest to do so. It should be noted that this is but one of a number of mechanisms that could be used to facilitate international cooperation. For example, various forms of bilateral joint implementation could accomplish the same objective. Hence, trade in emission rights is intended only as a proxy for any of a number of cooperative mechanisms.

With the above caveat in mind, we now describe our four cases:

- Case 1 (no interregional or intertemporal efficiency) -- Each OECD region is required to meet its annual emissions constraint independently. There is no trade in emission rights between the OECD and other regions.[20]

- Case 1a (interregional efficiency) -- The constraint is still on year-by-year emissions, but trade in emission rights is now permitted between the OECD and other regions. Non-OECD countries are allowed to emit in each period up to the level of their emissions in Case 1. If they reduce their emissions below this level, they may benefit from the sale of the emission rights generated.

- Case 1b (intertemporal efficiency) -- Rather than a set of year-by-year emission limits, the constraint on emissions from each OECD region is expressed as an upper limit on its cumulative emissions. This allows for higher emissions in years where the cost of emissions abatement is highest. "Payback" must occur by 2050. There is no trade in emission rights between the OECD and non-OECD regions.

- Case 1c (interregional and intertemporal efficiency) -- The constraint is now on cumulative emissions at the global level. Both interregional and intertemporal trading is permitted. Emission rights are based on Case 1. As a result, reductions take place both where and when it is cheapest to do so.

Figure 3 shows costs for Case 1 discounted to 1990 at 5% per year. The constraint on carbon-emitting activities leads to a reallocation of resources, away from the pattern that is preferred in the absence of this limit and into potentially costly conservation activities and fuel substitution. Relative prices change as well. These forced adjustments lead to a reduction in economic performance, as measured by GDP or some other indicator depending on the model. The tighter the constraint the greater the effect.

Figure 3. Costs of 20% Cut in OECD Emissions by
2010 -- Case 1
(costs through 2050 discounted to 1990 at 5%)

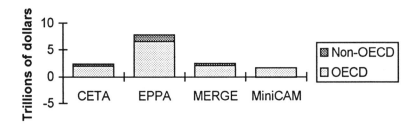

Note that, because of trade effects, many non-OECD countries will incur costs even when reductions are confined to the OECD. Restrictions on carbon emissions lead to lower OECD demand for oil, which results in lower revenue for the oil-exporting countries. In addition, an economic slowdown in the OECD countries affects the full range of developing country exports, and thus their growth. For many oil-importing developing countries these broader trade effects outweigh the gain from lower world oil prices. Three of the four models shown account for at least some of these effects (MiniCAM is the exception) and show a spillover of OECD loses onto non-OECD countries.

Not surprisingly, the models differ as to the magnitude of the economic impacts. This is to be expected given the large differences in emission baselines. EPPA, with the highest baseline, shows the highest costs. MiniCAM, with the lowest baseline, shows the lowest costs.

A second factor contributing to the large spread among models is the speed with which the capital stock is allowed to adjust to higher energy prices. As noted earlier, two of the models, EPPA and MERGE employ a so-called putty clay formulation. They attempt to track the economic lifetime of existing plant and equipment. As a result, these models show less responsiveness of energy demand to price changes in the short run than over the long run. Alternatively, models which assume greater malleability of capital (CETA and MiniCAM) produce lower cost estimates.

Potential gains from interregional efficiency. The models are in more agreement on the relative costs of the various alternatives (Figure 4). Note that the potential benefits from economic efficiency are substantial. In Case 1, each OECD region is required to act independently to reduce its emissions. There is no opportunity to take advantage of low-cost emission reduction options elsewhere in the world. From the perspective of global economic efficiency, this makes little sense. Clearly, it is inefficient to incur high marginal domestic abatement costs when low-cost alternatives exist in other countries. In Case 1a, we allow OECD countries to take advantage of the lower cost alternatives. We do this by permitting trade in carbon emission rights. Note that cooperation of this type can cut the costs of a carbon constraint by well over one-half.

Figure 4. Global Costs under Four Alternative Cases

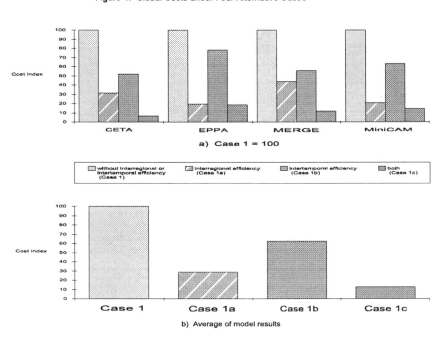

Figure 5 shows the impact on non-OECD countries. International cooperation not only reduces costs within the OECD, it may also result in substantial wealth transfers. Indeed, for three of our models, the revenue received from the sale of emission rights more than offsets the trade-related losses to non-OECD countries. Alternatively, one could devise a burden sharing scheme which imposes zero net costs on non-OECD countries.[21] Such a scheme would compensate non-OECD countries for losses accruing through international trade, but results in no additional wealth transfers. In this instance, costs to the OECD would be equivalent to global costs.

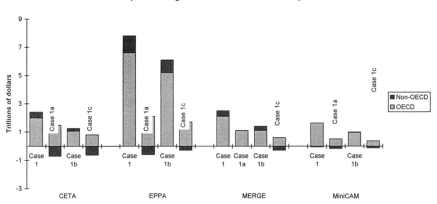

Figure 5. Regional Costs under Four Alternative Cases
(costs through 2050 discounted to 1990 at 5%)

Potential gains from intertemporal efficiency. We next turn to the issue of timing (Case 1b). When given the choice, each model shifts some emission reductions into the future. That is, it chooses to emit more in the early years with payback coming later on (see Figure 6a). This behavior can best be understood in terms of an optimal allocation problem. A constraint on cumulative emissions defines a carbon budget. That is, it specifies a total amount of carbon to be emitted over a fixed period of time. For Case 1b, each OECD region's carbon budget is defined as the sum of its permissible emissions between 2000 and 2050 (as specified in Case 1). The issue is how best to allocate the carbon budget over this period.

There are several factors that argue for using more of the available budget in the early years.[22] Deferring emission reductions provides valuable time to reoptimize the capital stock. Energy producing and energy using investments are typically long-lived (e.g., power plants, houses, transport). They were put into place with a particular set of expectations about the future. Abrupt changes are apt to be expensive. This is particularly the case when it comes to premature retirement of existing plant and equipment. Time is needed for the capital stock to adapt.

The optimal timing of emission reductions is also influenced by the prospects for new supply and conservation technologies. There has been substantial progress in lowering the costs of carbon-free substitutes (e.g., solar, biomass, energy efficiency) in the past. With a sustained commitment to R&D, there should be further cost reductions in the coming decades. It would make sense to draw more heavily on the carbon budget in the early years when the marginal costs of emissions abatement are highest. With cheaper alternatives in the future, there will be less need for reliance on carbon-intensive fossil fuels.

Figure 6. Global Emissions and CO2 Concentrations with and without
Intertemporal Efficiency (based on average of model results)

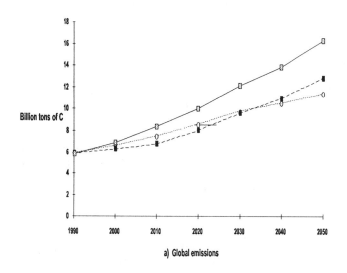

a) Global emissions

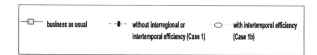

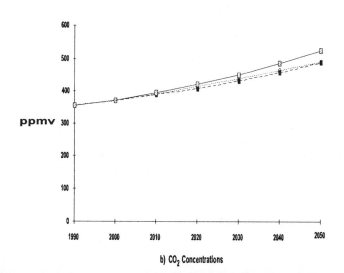

b) CO₂ Concentrations

Before leaving the timing issue, several additional caveats are in order. First, it should be noted that the two emission paths of Figure 6 result in different levels of atmospheric concentrations (prior to 2050). They may therefore differ in terms of environmental impacts. Given that the concentration paths lie so close together, however, the differential impacts on temperature and sea level are likely to be negligible.[23]

Second, the above considerations (capital stock turn over, R&D and discounting) argue for shifting some emission reductions into the future. They cannot, however, be used as an excuse for deferring these reductions indefinitely. The carbon budget is finite. There is an upper limit on the amount to be emitted between now and 2050 which continued deferral would soon exceed. The issue is one of optimal timing.

Finally, note that the amount of deferral depends on the size of the carbon budget. In this instance, there is insufficient flexibility to defer emission reductions altogether in the early years. The optimal emissions path lies between Case 1 and business as usual.

Returning to Figure 4, we see that the most efficient strategy is one which combines international cooperation with flexible timing (Case 1c).[24] In this instance, costs are reduced by more than 80%. Figure 7 provides some insight into why the savings are so large. It shows OECD GDP losses averaged across the four models. In Case 1, GDP losses grow to 2.4% over the next quarter century -- roughly $400 billion in today's economy. In Case 1b, GDP losses grow more slowly. Although annual losses exceed those of Case 1 toward the end of the time horizon, they are considerably lower early on. As a result, cumulative losses are smaller. If OECD countries are able to take advantage of low-cost emission reduction options elsewhere in the world, losses can be held to under 1% of GDP.

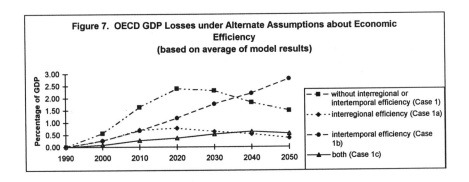

Figure 7. OECD GDP Losses under Alternate Assumptions about Economic Efficiency (based on average of model results)

The costs of less stringent carbon constraints. One way to reduce costs would be to design more cost-effective strategies. A second way would be to make the constraint less stringent. We now consider two additional variants on Case 1. In Case 2, we delay the date by which OECD countries must achieve the 20% reduction by 10 years. In Case 3, we put off the 20% reduction altogether. That is, OECD countries continue to hold emissions at 1990 levels.

From Figure 8, note that a substantial fraction of the costs of a 20% reduction would be incurred simply by extending the existing target. That is, much of the costs result from reducing emissions from the business-as-usual path to 1990 levels. Between 40% and 70% of the costs are associated with the decision to stabilize emissions at 1990 levels.

Figure 8. Costs of Alternative Sets of Targets and Timetables

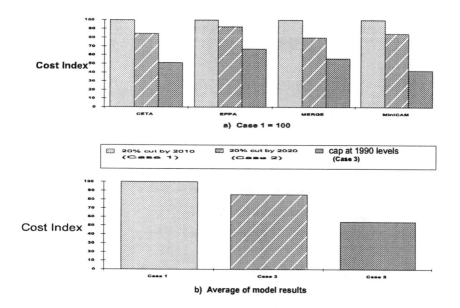

Figure 9 compares OECD GDP losses for the three cases. In Case 1, annual losses rise to 2.4% of GDP by 2020. Postponing the 20% cut by 10 years results in lower GDP losses during the initial two decades of the next century. But losses are similar thereafter. For Case 3, GDP losses are lower for the entire period. On average, lowering the target cuts GDP losses by nearly one-half.

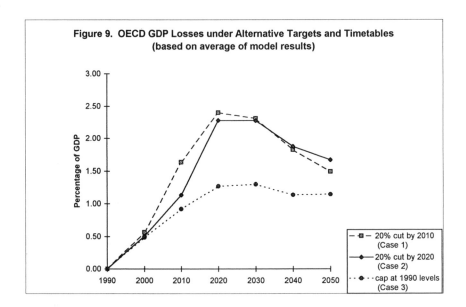

Figure 9. OECD GDP Losses under Alternative Targets and Timetables (based on average of model results)

5. Some Final Comments

Estimating mitigation costs is a daunting task. It is difficult enough to envisage the evolution of the energy-economic system over the next decade. Projections involving a half century or more must be treated with considerable caution. Nevertheless, we believe that exercises like the present one contain useful information. The value, however, lies not in the specific numbers, but in the insights for policy making. With this in mind, we attempt to summarize what we have learned.

- Implementing an AOSIS-type proposal may require substantial CO2 reductions for OECD countries. With a growing emissions baseline, more and more carbon must be removed from the energy system to maintain an absolute target. Such reductions could be quite costly -- perhaps, as much as several percent of GDP to OECD countries.

- Because of trade effects, the non-OECD countries likely will incur costs even when emissions reductions are confined to the OECD. Restrictions on carbon emissions lead to lower demand for oil, which results in lower revenue for oil-exporting countries. In addition, an economic slowdown in the OECD countries affects the full range of developing country exports, and thus their growth. For many oil-importing developing countries, these broader trade effects outweigh the gain from lower world oil prices.

- One way to reduce mitigation costs would be to design cost-effective constraints. Indeed, the present analysis suggests that the potential gains from international cooperation (interregional efficiency) and flexible timing (intertemporal efficiency) are huge. Taken together, they can reduce costs by more than 80 percent. The key is to allow emission reductions to take place both *where* and *when* it is cheapest to do so.

- A second way to reduce mitigation costs would be to adopt less stringent constraints. For example, rather than a 20 percent cutback, the OECD could agree to hold emissions at 1990 levels. The analysis suggests that the reduction in overall mitigation costs would be between 30 and 60 percent. The savings, however, must be weighed against the impacts of the incremental emissions through larger changes in climate.

- The following steps could substantially reduce the costs of implementing a carbon constraint under the Berlin Mandate: 1) allow developed countries to purchase low-cost abatement options in developing countries, 2) allow time for the economic turn over of existing plant and equipment, 3) invest in the development of economically attractive substitutes for carbon-intensive fuels, and 4) ensure that cost-effective options are adopted to the greatest extent possible.

- Our results are consistent with other studies which suggest that carbon emissions will continue to grow in the absence of policy intervention. Proposals which focus exclusively on developed countries may slow the growth in global emissions, but they will not stabilize them at anywhere near present levels. Nor will they stabilize atmospheric concentrations, the ultimate goal of the Framework Convention. To do so, would eventually require developing country participation.

The present paper identifies enormous savings from international cooperation and flexible timing. Realizing this potential, however, may be another matter. For example, how do we divide up the savings from international cooperation? Or, how do we ensure that parties maintain a credible path toward fulfilling commitments? Considerable ingenuity will be required, but given the stakes, even partial success is likely to be well worth the effort.

Fortunately, some of the necessary concepts are already being tested. For example, efforts to incorporate international cooperation can build upon the experience gained from national and international joint implementation initiatives. With regard to flexible timing, a limit might be placed on a country's cumulative emissions. Subject to this constraint, the country could lay out its own projected emissions time path and prepare a formal plan that builds on existing experience with National Action Plans under the Framework Convention. Periodic reviews could then track adherence to the commitment. Technology development efforts, with suitable performance milestones, also could be an integral part of both the path definition and review processes.

Negotiators must consider a myriad of competing ideas and interests inherent in shaping a global policy. One of their greatest challenges will be to meet the injunction of Article 3 of the Framework Convention: "policies and measures to deal with climate change should be cost-effective so as to ensure global benefits at the lowest possible costs." Our success in confronting the challenge of climate change may depend directly on their success in doing so.

The larger question, of course, is what constitutes an appropriate set of emission constraints. This requires consideration of both benefits and costs. The present analysis has been confined to the cost side of the ledger. That is, we examine the costs of reducing CO2 emissions. Policy makers will also want to know what they are buying, in terms of reducing the undesirable consequences of global warming. Such an analysis is beyond the scope of the present effort.

Endnotes

[1]With contributions from Henry Jacoby (Massachusetts Institute of Technology), Alan Manne (Stanford University), Stephen Peck (Electric Power Research Institute), Tom Teisberg (Teisberg and Associates), Marshal Wise (Pacific Northwest Laboratory) and Zili Yang (Massachusetts Institute of Technology). This paper reports initial results of the Subgroup on the Regional Distribution of the Costs and Benefits of Climate Change Policy Proposals, Energy Modeling Forum-14, Stanford University. Helpful comments were received from Sally Kane and John Weyant. The authors are much indebted to Amy Craft for research assistance. The views presented here are solely those of the authors.

[2]For the text of the Berlin Mandate, see United Nations Climate Change Bulletin, Issue 7, 2nd Quarter 1995, published by the interim secretariat for the UN Climate Change, Convention, Geneva. For the text of the Framework Convention, see Intergovernmental Negotiating Committee for A Framework Convention on Climate Change, Fifth Session, Second Part, New York, 30 April-9 May 1992.

[3] See Intergovernmental Panel on Climate Change (IPCC), Report of Working Group III, Chapter 1, forthcoming, Cambridge University Press.

[4]The four models comprise the Subgroup on the Regional Distribution of Costs and Benefits of Climate Change Policy Proposals. The Subgroup is open to models participating in Stanford University's Energy Modeling Forum (EMF) Study on "Integrated Assessment of Climate Change."

[5] See Intergovernmental Negotiating Committee for A Framework Convention on Climate Change, Fifth Session, Second Part, New York, 30 April-9 May 1992.

[6]See Intergovernmental Panel on Climate Change (IPCC), Report of Working Group III, Chapter 9, forthcoming, Cambridge University Press.

[7]See Intergovernmental Panel on Climate Change (IPCC). *Climate Change 1994,* Cambridge University Press, 1994. Also see, Wigley, T., Richels, R. and Edmonds, J. "Economic and Environmental Choices in the Stabilization of Atmospheric CO2 Concentrations," *Nature*, Vol. 379, 18 January, 1996.

[8]See Intergovernmental Panel on Climate Change (IPCC), Report of Working Group III, Chapters 9 & 10, forthcoming, Cambridge University Press.

[9]See Peck, S. and Teisberg, T. "International CO2 Emissions Targets and Timetables: Analysis Using CETA-M", Working Paper, 1995.

[10]See Z. Yang, et. al, "The MIT Emissions Projection and Policy Assessment (EPPA) Model", Draft report, MIT Joint Program on the Science and Policy of Global Change, February 1996.

[11]See Manne, A. and Richels, R. "The Berlin Mandate: the Costs of Meeting Post-2000 Targets and Timetables", Stanford University, Stanford , CA, forthcoming in *Energy Policy*, 1995.

[12]See Edmonds et al

[13]See Hogan, W. and Jorgenson, D. "Productivity Trends and the Costs of Reducing CO2 Emissions'" *Energy Journal* 12 No. 1, 1991.

[14]For a detailed model comparison, see EMF-14.

[15]These projections are intended as examples of how emissions might evolve under existing policies. They should not be interpreted as each analysis team's "best guess" of future emissions.

[16]See Intergovernmental Panel on Climate Change (IPCC). *Climate Change 1994,* Cambridge University Press, 1994.

[17]See Manne, A. and Richels, R. "The Costs of Stabilizing Greenhouse Gas Emissions: A Probabilistic Analysis based on Expert Judgments," *The Energy Journal* 15(1), 1994.

[18]The AOSIS proposal calls for Annex 1 countries to reduce emissions by 20% by 2005.

[19]See notes 15 and 16.

[20]There is some trade in emission rights within the OECD, however. This is the consequence of aggregating single countries into larger regions.

[21]With an international market in carbon emission rights, global abatement costs are independent of the burden sharing scheme. This allows us to separate the difficult issues of efficiency and equity. For the theoretical considerations underlying this proposition, see Manne, A. "Greenhouse Gas Abatement - toward Pareto Optimality in Integrated Assessments", in Education in a Research University, edited by Kenneth J. Arrow, Richard W. Cottle, B. Curtis Eaves and Ingram Olkin, Stanford University Press, Stanford CA, 1996

[22]For a more detailed discussion of the timing issue, see Wigley et al, note 16.

[23]For the analysis, we use the carbon cycle model of Wigley. See Wigley, T.M. "Balancing the Carbon Budget: the Implications for Projections of Future Carbon Dioxide Concentration Changes," *Tellus*, 45B, 1993.

[24]EPPA is a recursive rather than an intertemporal optimization model. Several alternative emission paths were explored for Cases 1b and 1c. The results reported here are for the lowest-cost of the paths tested, and the results are not strictly comparable with those from the other models.

Chapter 6

SOME IMPLICATIONS OF IMPROVED CO$_2$ EMISSIONS CONTROL TECHNOLOGY IN THE CONTEXT OF GLOBAL CLIMATE CHANGE

Stephen C. Peck
Electric Power Research Institute
Palo Alto, California 94303
Thomas J. Teisberg
Teisberg and Associates
Charlottesville, Virginia 22901

Abstract: In this paper, we use the CETA-R model to explore some implications of improved (i.e. lower cost) CO2 control technology. CETA-R represents the costs of global climate change in terms of small risks of large climate change related losses.

First, we treat technology cost as deterministic and explore the implications of improved control technology for the value of information about the risks of large climate change related losses. We find that when future control technology is better, the value of information about climate change risks is substantially lower.

Second, we treat future technology as uncertain and explore the value of information about it, and the economic benefits of research and development that may produce improvements in it. We find that the value of information regarding future technology is relatively low. On the other hand, the direct benefit of better technology is large -- each percentage point increase in the chance of twenty percent improved technology has a present value of $8 to $12 billion, depending on the magnitude of the risk of large climate change related losses.

1. Introduction

Since the beginning of the industrial revolution, emissions of greenhouse gases, such as CO2, have been accumulating in the earth's atmosphere at a steady rate. Because these gases enhance the natural heat trapping ability of the earth's atmosphere, many people fear that the accumulation of greenhouse gases will cause the earth's climate to

John Weyant (ed.), ENERGY AND ENVIRONMENTAL POLICY MODELING. Copyright © 1998. Kluwer Academic Publishers. ISBN 0-7923-8348-6. All rights reserved.

change in undesirable ways. However, there is still great uncertainty about the degree to which climate may change, and the extent to which the consequences of climate change will prove harmful.

The time scale of global climate change is unusually long. Climate change is driven by cumulative emissions of CO_2 and other greenhouse gases over centuries, and the full climate response to a given level of cumulative emissions is delayed by at least a few decades, due to the thermal inertia of the oceans.

On this time scale of years to decades, significant improvements in CO_2 emissions control technology are possible. In this paper, we use the CETA-R model to explore some implications of improved technology. Specifically, we explore the implications of improved technology for the value of information about the risks of large climate related losses. Also, we explore the value of information about future technology costs, and we examine the direct benefits from improving the odds of lower cost CO2 control technology through research and development.

We find that when future control technology is better, the value of information about climate change risks is substantially lower. In addition, we find that the value of information regarding future technology is relatively low. On the other hand, the direct economic benefit of better technology is large.

2. The CETA-R Model

CETA-R is a small scale "integrated assessment" model of the climate change problem, from the perspective of the world as a whole. The model combines simple representations of the climate system and the economic system, thereby providing a consistent framework for analyzing alternative policies that might be adopted to limit or slow climate change. An optimal solution of the model is a time path of CO_2 emissions that achieves an appropriate balance between the costs of global warming and the costs of measures that slow this warming.

The following sections briefly describe the economic and climate systems in CETA-R, and how costs of climate change are modeled in CETA-R.[1]

2.1 Economic System

Economic output in CETA-R is a function of inputs of labor, capital, electric energy, and non-electric energy. Labor input and the initial capital stock are exogenously specified. Future capital depends on investment decisions, which are among the key endogenous variables in the model.

Choices about electric and non-electric energy inputs to production are also key endogenous variables of CETA-R. The amount of energy employed in producing output, and the energy technologies used, together determine the costs incurred in energy production and the CO2 emissions generated as a result of energy production.

Besides emissions of C02, we also incorporate exogenously specified emissions of CH4, N2O, and CFCs. Initially, emissions of CH4 and N2O are assumed to be at levels roughly consistent with IPCC Scenario B projections. However, since CO_2 emissions are virtually eliminated by 2200 in all of the cases we consider, we assume that emissions of the other greenhouse gases are phased out as well. Thus, emissions of CH4 and N2O are eliminataed between 2010 and 2110, and CFC emissions are eliminated immediately.

2.2 Climate System

The climate system in CETA and CETA-R contains very simple representations of the behavior of greenhouse gases in the atmosphere, equilibrium warming, and the lag of actual warming behind equilibrium warming.

We represent the removal of CO2 from the atmonsphere using the simple CO_2 impulse response function developed by Maier-Reimer and Hasselmann (1987).[2] Behavior of the othe rexogenous greenhouse gases is represented using a trnsition matrix presented in Nordhaus (1990) that governs the removal or transformation of these gases over time.

Equilibrium warming from CO2 is represented as proportional to the logarighm of CO2 concentration relative to pre-industrial concentration, with a constant of proportionality that is calibrated to climate sensitivity results from general circulation model (GCM) equilibrium experiments.[3] Equilibrium warming from the other greenhouse gases is a linear function of the square root of concentration for CH4 and N2O, and it is linear in concentration for CFCs.[4]

Thermal lag is represented by a simple geometric adjustment process, which is calibrated to characteristic response times (e-fold times) obtained in experiments with coupled ocean-atmosphere models.[5]

2.3 Uncertain Losses from Temperature Rise

In CETA-R, the costs of climate change are large welfare losses which may occur with relatively small probabilities that depend on the extent of temperature rise.[6] To impliment this approach, we draw upon some expert survey results reported in Nordhaus (1994). Nordhaus's results provide estimates of the probabilities of warming-related losses over the next century or two. We use these estimates to calibrate loss probability functions which specify the probabilities of a large welfare loss as a function of temperature rise, in each 20-year period of our model.

Nordhaus surveyed about 20 experts in economics and the natural sciences regarding their subjective probabilities of damage from global warming. Among other things, the experts were asked to assess the probability of a "high-consequence outcome" -- one defined as a lowering of global incomes by 25 percent or more (the economic equivalent of the Great Depression)."[7] Nordhaus found that the experts who are natural scientists' probabilities of a high-consequence event are roughly an order of magnitude greater than the probabilities of the group as a whole.

Given the great difference of opinion between the natural scientists and the experts as a whole, we calibrate loss probability functions separately to match the survey responses of the scientists and the experts as a whole. These two calibrations provide natural "high" and "Low" cases for the costs of climate change. The resulting calibrated loss probability functions are shown in Figure 1; these functions give the probability of a 25 percent welfare loss occurring in a given 20-year model time period, for temperature rises ranging from .5 degrees C. to 5.5 degrees C.[8]

Figure 1: Calibrated Loss Probability Functions

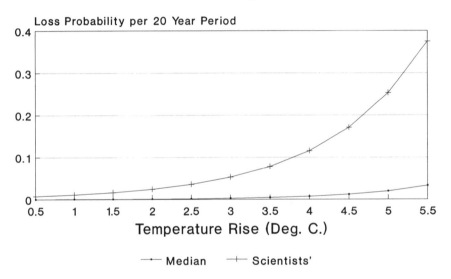

3. Technology Assumptions

Technology in CETA-R is represented by three key parameters. These are the costs of the electric and non-electric backstop technologies, and the Autonomous Energy Efficiency Improvement (AEEI) rate. In the present analysis, we treat these parameters as together defining the state of technology.

The backstop technologies are admittedly speculative future technologies that are characterized as providing carbon-free energy in practically unlimited amounts, but at relatively high costs, and not until some later date. The electric backstop technology might be photovoltaic or some form of advance nuclear power. The non-electric backstop technology might be hydrogen produced by electrolysis, where the required electricity is obtained using the electric backstop technology. In the CETA-R base case, the cost of the non-electric backstop is about $90 per barrel of oil equivalent.

The AEEI is a parameter that governs how much energy inputs may be reduced over time without reducing output, for tiven inputs of labor and capital. In recent years, there has been a trend towerd reduced energy inputs per unit of output, even in the face of flat to declining real energy prices. We use the AEEI parameter to project this trend into the future. In the CETA-R base case, the AEEI parameter is .68 percent per year, implying that energy input may fall at this rate without any effect on output, ceteris paribus.[9]

Our base case values of the electric and non-electric backstop costs and the AEEI parameters define our "Standard Technology" case. For our "Improved Technology" case, we reduce the cost of the electric and non-electric backstop technologies by 20 percent and we raise the Autonomous Energy Efficiency Improvement rate by 20 percent, starting in 2050.

4. Implications of Loss Probability and Technology Assumptions

We begin our analysis by exploring the sensitivity of time paths of optimal carbon emissions and optimal carbon shadow prices to assumptions about the loss probability function and the state of technology. For our loss probability assumptions, our "High Damage" case uses the loss probability function calibrated to the scientists' loss estimates. Our "Low Damage" case uses the loss probability function calibrated to the median estimates. For our technology assumptions, we use the Standard Technology and Improved Technology cases as described above.

Figure 2 shows the sensitivity of the optimal carbon emissions paths to the loss probability and technology assumptions. When damage is High, the optimal paths are much lower than they are when damage is Low; this is particularly true beginning after 2050, when the backstop technologies start to become available in significant quantities. For either assumption about damage, the technology assumption makes a modest difference in the optimal path -- specifically, when technology is better, the optimal path is somewhat lower. It is interesting, however, that the technology assumption affects optimal emissions more strongly when damage is Low than when it is High.

To understand what underlies the optimal emissions paths in Figure 2, we next present a couple of figures which show the energy technologies that are employed across two of the four scenarios represented in Figure 2. Figure 3, for example, shows technologies employed in the Low Damage - Standard Technology case, which is the one with the highest optimal emissions path. Areas in the figure represent energy production (and consumption) by technology used, and the areas cumulate up to the total amount of energy produced.

Figure 2: Optimal Carbon Emissions

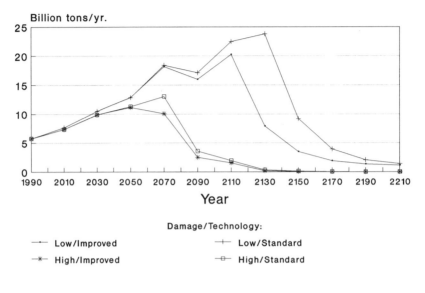

Damage/Technology:

⟋ Low/Improved   + Low/Standard

✳ High/Improved   ⊟ High/Standard

Figure 3: Energy Use
Low Damage - Std. Technology

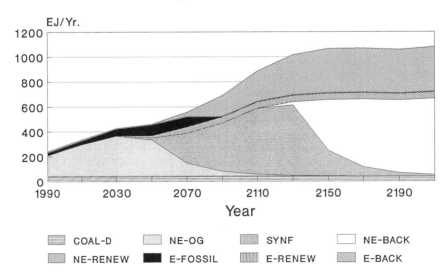

COAL-D   NE-OG   SYNF   NE-BACK

NE-RENEW   E-FOSSIL   E-RENEW   E-BACK

Following are brief descriptions of each of the areas shown in Figure 3, beginning at the bottom of the Figure:

(1) COAL-D, direct use of coal, continues at a low rate throughout the time horizon shown.

(2) NE-OG, non-electric oil and gas use, rises to a peak around 2030, before declining to near zero by 2110. Oil and gas is an inexpensive technology and relatively low in carbon (compared to coal, at least); thus it is fully used until the resource base can no longer support it.

3) SYNF, liquid synthetic fuels made from coal, comes on-stream about 2070 to replace the declining production of oil and gas. Synthetic fuels are very carbon intensive, and their production underlies the bulge in emissions observed in Figure 2 between 2070 and 2170. In the Low Damage - Standard Technology case represented in Figure 3, synthetic fuels are produced until the coal resource base is exhausted, in spite of the large carbon emissions that result.

(4) NE-BACK, the non-electric backstop technology, comes on-stream after 2130 to replace the declining production of synthetic fuels.
(5) NE-RENEW, non-electric renewables, begin to be used immediately after their introduction date in 2050.

(6) E-FOSSIL, electricity produced from fossil fuels, continues until about 2090.

(7) E-RENEW, hydroelectric and geothermal electric, is used at capacity throughout the time horizon shown.

(8) E-BACK, the electric backstop technology, comes on-stream after 2050, when it first becomes available in quantity. It completely replaces electricity derived from fossil fuel technologies by 2090; the temporary dip in emissions in 2090 (see Figure 2) is caused by this relatively abrupt replacement of fossil fuels by the electric backstop.

Figure 4 shows energy technologies used in the Low Damage - Improved Technology scenario, which produces the next highest optimal emissions path in Figure 2. The main difference between Figure 4 and Figure 3 is that synthetic fuels are phased-out a little earlier in Figure 4. This is what accounts for the somewhat earlier and lower peak in optimal emissions in this scenario.

While we do not show energy use figures for the High damage cases, energy use in these cases is similar, except that synthetic fuels use is very small in the High Damage - Standard Technology case, while it is virtually eliminated in the High Damage - Improved Technology case. Since there is so little synthetic fuels production in either of these High damage cases, Improved technology tends to have only a modest effect on optimal emissions in these cases. This is in contrast to the Low damage cases, in which there is more synthetic fuels use to be displaced if technology is Improved.

Figure 4: Energy Use
Low Damage - Improved Technology

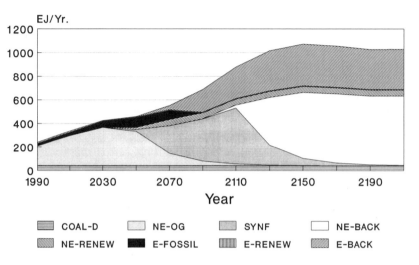

| COAL-D | NE-OG | SYNF | NE-BACK |
| NE-RENEW | E-FOSSIL | E-RENEW | E-BACK |

The preceding examination of energy use across scenarios suggests a simple summary interpretation of the optimal emissions paths in Figure 2. Across all scenarios considered, oil and gas is used in the first half of the next century; this is what accounts for the high degree of similarity in the optimal emissions paths in this early time frame. After about 2050, however, there are big differences in the extent to which synthetic fuels are used. When they are used heavily, as in the Low Damage scenarios, optimal emissions rise sharply to a new peak around 2110 or 2130; but when synthetic fuels are used lightly or not at all, as in the High Damage scenarios, optimal emissions begin to fall as oil and gas use tapers off.

Lastly, Figure 5 reports the sensitivity of optimal carbon shadow prices to the damage and technology assumptions (for clarity, prices above about $800 per ton have been omitted from the Figure).[10] Optimal carbon shadow prices measure both the marginal cost of emissions control and the marginal present value expected future damages from emissions. As was true for the optimal carbon emissions paths, the optimal carbon shadow price paths are more strongly affected by the damage assumption than by the technology assumption. Also, as was true for the emissions paths, the technology assumption has a smaller effect on the carbon shadow price path when damage is High. This, again, is because there is little synthetic fuels production to be displaced in the two High damage cases, and thus there is little difference in emissions between these cases; little difference in emissions implies that there is little difference in marginal present value expected future damages (i.e. in optimal carbon shadow prices).

Figure 5: Carbon Shadow Prices

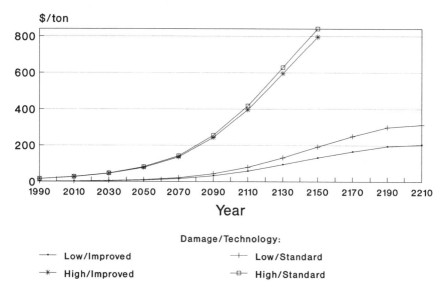

The preceding sensitivity results indicate that the climate change damage assumption is the more important determinant of optimal emissions. Still, the technology cost assumption has an important affect on optimal emissions, even for the 20 percent technology improvement that we have assumed. While the effect of technology is small before about 2070, after 2070, optimal emissions are significantly lower if technology is better.

Since climate change costs depend on cumulative emissions over decades and climate response to emissions is slow, it is reasonable to hypothesize that lower optimal future emissions would be accompanied by higher optimal near term emissions. However, differences in optimal near term emissions, if any, are too small to be discernible in Figure 2. Thus, to test the above hypothesis, Table 1 shows optimal emissions in 2030 and the difference in emissions due to improved future technology, for both climate change cost assumptions. The numbers in the table show that there is an increase in near term emissions when future technology is better, thus confirming our hypothesis.

The reason that the change in near term optimal emissions in Table 1 is small is that optimal emissions in the near term are generally insensitive to underlying assumptions. As we saw earlier, this insensitivity is attributable to the fact that conventional oil and gas is the key energy technology in this time frame, and it is optimal to make use of the oil and gas technology, whether climate change costs are "High" or "Low," as we have defined these scenarios here.

Table 1

Optimal Emissions in 2030
And the Increase Due to Improved Future Technology

Future Technology	Low Damage	High Damage
Improved	10.510	9.897
Standard	10.496	9.881
	-----	-----
Increase	.014	.016

5. Improved Technology and the Value of Information about Damage

In this section, we treat damage (High or Low) as uncertain, while the state of technology is assumed to be known and to either be Standard technology or Improved technology. Under these assumptions, we estimate the value of information regarding the damage. Specifically, we assume that the damage level becomes known either immediately or after 2030. The difference in expected utility between these two cases can then be used to obtain a measure of the value of perfect information now, versus after 2030. We calculate these measures over a range of assumptions about the prior probability that damage will be High, as well as for the two possible states of technology.

Figure 6 graphically displays our results. The most notable feature of these results is that when future technology is better, the value of information about damages is lower. The intuition for these results is as follows. Better future technology means that it will be easier to reduce emissions in the future. This means that control decisions in the early decades (prior to 2050) are somewhat less critical, which in turn means that information early in this time frame is less valuable.

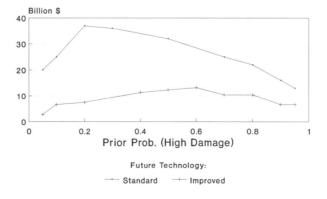

Figure 6: Value of Information About Damage

Finally, Figure 6 reveals a subtle difference in the behavior of the value of information about damage as a function of prior probability of High damage. This is the tendency for the value of information to peak at a much lower probability in the Standard technology case than in the Improved technology case. The intuition for this result is as follows.

The value of information is typically highest at the prior probability where one is just indifferent between two possible actions -- for example, a "little more" control or a "little less" control, in the near term. When future technology will be high cost (i.e. Standard), it will be more expensive to "fix" excessive early emissions by controlling more in the future, and one is indifferent between extra near term control and no extra near term control at a relatively low probability of High damage. Conversely, when future technology will be low cost (i.e. Improved), it will be less expensive to "fix" excessive early emissions by controlling more in the future, and one is indifferen" between extra near term control and no extra near term control at a relatively high probability of High damage. Thus, the value of information peaks at a relatively low probability of High damage (.2) in the Standard technology case, and at a relatively high probability of High damage (.6) in the Improved technology case.

6. Value of Information about Technology and Benefits of Improved Technology

In this section, we explore the value of information about technology, and the direct value of Improved technology. Our approach to valuing information about technology is analogous to that we used in the preceding section. That is, we treat future technology as uncertain (it may be Improved or Standard, as defined earlier), while damage is assumed to be known and either High or Low. Under these assumptions, we estimate the value of early information regarding future technology. As before, we calculate values over a range of assumptions about the prior probability that technology will be Standard, as well as for the two possible assumptions about damage.

Figure 7 displays the value of information results, using the same vertical scale as was used in Figure 6. In general, the value of information about future technology is relatively low, and not particularly sensitive to the underlying damage assumption.

This result is not surprising in light of the carbon shadow price paths reported in Figure 5. Those paths indicate that the technology assumption has a much smaller effect on optimal carbon shadow prices than the damage assumption. Since the carbon shadow price is an indicator of the intensity of control effort, the smaller effect of the technology assumption implies that optimal control effort is less dependent on the technology assumption than it is on the damage assumption. Under these circumstances, it is reasonable to find that the value of information about technology is lower than the value of information about damage.

Figure 7: Value of Information About Technology

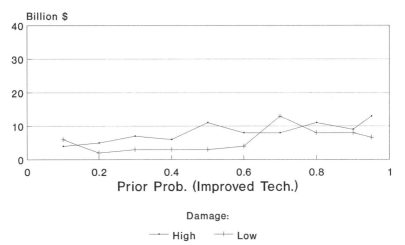

Billion $

Damage:

—•— High   —+— Low

Figure 8 displays the direct value of improved technology. The figure shows the economic benefit of increasing the probability of improved future technology by 10 percentage points, as a function of the initial probability of improved technology. Evidently, this benefit is not dependent on the initial probability, though it does depend on the damage assumption in a plausible way -- i.e. when damage is High, the benefit of Improved technology is greater. Averaging these benefits over the initial probabilities at which they were calculated, the direct benefit of Improved technology is about $8 billion per percentage point when damage is Low, and $12 billion per percentage point when damage is High.

Figure 8: Benefit of 10 Percentage Point Higher Prob. of Improved Tech.

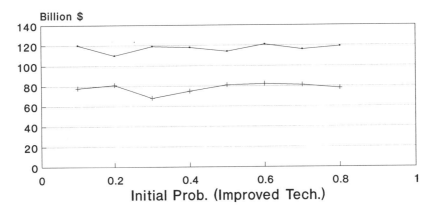

Billion $

Damage:

—•— High   —+— Low

7. Summary and Conclusions

In this paper, we use CETA-R, a small scale "integrated assessment" model of climate change, to explore the role that technological improvement plays in the context of the global climate change issue. We consider (1) the implications of improved technology for the value of information about damages, (2) the value of information about technology itself, and (3) the direct economic benefits of increasing the odds of improved technology.

We find that when future technology is better, the value of information about damages is substantially lower. We also find that the value of information about technology is relatively low, but that the direct economic benefit of improving future technology is very high.

These results suggest that research and development aimed at improving future technology has a high payoff, both directly and indirectly. The direct payoff is the economic benefit of improved technology. The indirect payoff is a reduced urgency to resolve uncertainties about the damages from global climate change. In effect, research to lower the costs of future technology can be viewed in part as a hedge against possible failure to resolve uncertainties about climate change damages, in advance of those damages actually manifesting themselves.

References

Chao, Hung-Po (1995), "Managing the Risk of Global Climate Catastrophe: An Uncertainty Analysis," *Risk Analysis*, Vol. 15, No. 1.

Maier-Reimer, E. and K. Hasselmann (1987), "Transport and Storage of CO2 in the Ocean -- An Inorganic Ocean-Circulation Carbon Cycle Model", *Climate Dynamics*, Vol. 2, pp. 63-90.

Manne, Alan S. and Richard G. Richels (1994), "The Costs of Stabilizing Global CO2 Emissions: A Probabilistic Analysis Based on Expert Judgements," *The Energy Journal*, Vol. 15, No. 1, pp. 31-56.

Nordhaus, William D. (1990), "Contribution of Different Greenhouse Gases to Global Warming: A New Technique for Measuring Impact," February 11.

Nordhaus, William D. (1994), "Expert Opinion on Climatic Change," *American Scientist*, January-February, pp. 45-51.

Peck, Stephen C. and Thomas J. Teisberg (1995), "Optimal CO2 Control Policy with Stochastic Losses from Temperature Rise," *Climatic Change*, Vol. 31, No. 1, September, pp. 19-34.

Schlesinger, Michael E. and Xingjian Jiang (1990), "Simple Model Representation of Atmosphere-Ocean GCMs and Estimation of the Time Scale of CO2-Induced Climate Change," *Journal of Climate*, December, pp. 1297-1315.

Endnotes

[1] For a more extended description of CETA-R, see Peck and Teisberg (1995).

[2] The Maier-Reimer and Hasselmann model represents CO2 uptake by the ocean. We assume that the biosphere is neutral, i.e. it has no net CO2 uptake.

[3] Sulfur aerosols have recently been identified as an important offset to the warming otherwise expected from CO2. We omit this effect because we lack a simple way to represent it in our model. Since sulfur aerosols have a very short atmospheric residence time, and since these aerosols are a function of fossil fuel use, the effect of omitting them from our model is that temperature rise is overstated during the years before fossil fuel use ends.

[4] Although we treat CFCs as a greenhouse gas, recent scientific work calls this assumption into question, since CFCs interact with lower stratospheric ozone in a way that may largely offset the direct radiative forcing of the CFCs. To test the practical importance of this effect in our model, we have experimented with eliminating CFCs altogether, and found no significant effect on optimal carbon emissions.

[5] We assume an e-fold time of 50 years based on Schlesinger and Jiang (1990).

[6] See Chao (1995) for a steady-state analysis using this kind of approach to climate change costs.

[7] Nordhaus (1994), p.47.

[8] We assume that welfare losses persist indefinitely. At the opposite extreme, they might last one time period (20 years); in this case, optimal carbon shadow prices would be reduced by about 60 percent. It would be possible to implement the model with a probability of recovering a prior loss, though this is something we have not yet done.

[9] This figure is adopted from expert survey results reported in Manne and Richels (1994).

[10] In fact, for shadow prices over roughly $800 per ton, there is little additional emissions reduction forthcoming, since any newly installed (as opposed to pre-existing) energy technology will be carbon free at this shadow price. Shadow prices above this level may be interpreted as potential values of removing carbon from the atmosphere.

Chapter 7

DETERMINING AN OPTIMAL AFFORESTATION POLICY: A COST-BENEFIT ANALYSIS

Susan Swinehart
Stanford University

Abstract: *Afforestation, tree planting on land that otherwise would be fallow, is being discussed as a method for reducing net carbon emissions. This article summarizes Swinehart (1996), which presents a tool for determining the optimal level of domestic afforestation. The tool is TREES, a model that evaluates possible afforestation projects in the U.S. We present an overview of the model structure and data. One of the crucial data parameters is the valuation of carbon. The international price of carbon, as computed by MERGE, provides TREES with a one-way linkage with a global integrated assessment (GIA) model. Sequestration results for scenarios utilizing different carbon valuations are presented and are compared with results from traditional abatement measures. The policy implications from these results are discussed.*

Section 1: Introduction

Motivation for Afforestation

Sequestration of carbon through afforestation to offset global warming is not a new concept. References to papers on increasing carbon sinks date from at least 1976. Yet most research efforts on greenhouse gas abatement have been concerned with reducing fossil fuel use, not on increasing biosphere uptake of carbon. The lack of investigation into afforestation can be explained in part by the uncertainties involved. For instance historical fossil fuel emissions for each country are generally known, and the release of CO_2 from fossil fuel burning is a deterministic chemical process: abatement activities and the CO_2 not emitted can be modeled directly. However the carbon content of a forested area varies greatly by region,

John Weyant (ed.), ENERGY AND ENVIRONMENTAL POLICY MODELING. Copyright © 1998. Kluwer Academic Publishers. ISBN 0-7923-8348-6. All rights reserved.

tree species, land treatment, and by the use of the land and any harvested wood thereafter.

The lack of extensive examination of afforestation can be attributed to additional causes. Not only is the amount of carbon sequestration unknown, but carbon sequestered by trees may be re-released into the atmosphere through risks such as disease and fire. With these uncertainties the effectiveness of money spent on afforestation is less clear than it would be with other mitigation methods. Because large scale planting efforts are not immediately implementable and initial growth rates for most tree species are low, carbon benefits are realized much later than are most of the project costs. Delays range from 10 to 40 years, a time scale comparable to replacing existing power plants with more efficient or less carbon intensive units at the end of their service lives.

Domestic sequestration efforts are constrained by availability of land and even at maximal land usage could offset only a fraction of current U.S. emissions. In addition high sequestration rates cannot be maintained indefinitely. A mature forest no longer actively absorbs carbon in net. In an analysis of the global potential for carbon storage in biomass, the GECR (1995) finds that the upper limit would entail a doubling of terrestrial carbon storage. One third of the carbon from known remaining fossil fuel reserves could be sequestered with this effort at most.

In light of these disadvantages we might question why afforestation should be considered as a mitigation option. A simple answer is that afforestation is another option available. Operations Research dictates that allowing for additional options improves the solution when optimizing a problem. In a cost-benefit analysis the optimal mitigation policy is also the least costly of all policies that satisfy a particular abatement path. This strategy includes the most inexpensive options from both the traditional methods of abatement and the afforestation projects. The marginal costs across each of these strategies should be equal. Researchers agree that some level of afforestation will be cost-effective; the disagreement is over what this level may be. Excluding afforestation from the analysis can only raise the costs.

Another reason is provided by examining a future scenario where carbon emissions from the energy sector are largely eliminated (as is indicated in the case of high sensitivity to climate change for MERGE by Manne and Richels (1995)). A further decrease in atmospheric carbon dioxide levels may still be desirable, yet traditional abatement methods are already constrained. Planting trees to absorb atmospheric carbon is a viable option to force *net* emissions negative. Afforestation is actually a geo-engineering method and is the only such method known to be effective and safe. Other geo-engineering possibilities, such as fertilizing the oceans with iron or launching dust particles into the atmosphere, have the potential to be cheaper. However they are not proven technologies and may have unforeseen consequences when employed on a large scale.

The use of afforestation to decrease atmospheric CO_2 levels has a certain appeal. Aside from mitigating global warming, growing trees may have additional benefits such as reducing erosion and providing timber resources. However many of the proposed afforestation schemes involve the creation of monocultural plantations rather than diverse and scenic forests. Therefore environmental groups may oppose rather than endorse large scale efforts. In addition to the direct costs associated with land procurement and tree planting, an afforestation program may have repercussions and incur further costs from interactions with the agricultural and timber sectors.

Afforestating and Halting Deforestation: Domestic verses Global Solutions

Most U.S. afforestation projects would involve land that was at one time forested, even if as far back as the pre-colonial days. That land is more likely to be suited to tree growth than is naturally unforested land. One exception may be in the upper latitudes where global warming might eventually extend the forest line north. The term *afforestation* is used when establishing tree stands on land not currently used for forestry. One definition is the planting trees of on land that has either remained unforested for at least 50 years or that is used for a non-timber purposes such as for agriculture. *Reforestation* refers to replanting lands currently designated for forestry. Although a mere distinction in semantics, we note that reforestation occurs on a regular basis in developed nations, whereas few large-scale afforestation projects have yet been realized. This definition needs to be clear in actual policy implementation because land owners have strong incentives for classifying existing timberland as afforestation project land.

In comparing reforestation or afforestation projects with efforts to halt deforestation, we should consider that replanting trees in a deforested area does not counteract the carbon loss from destroying the forest until a long period of time has passed. Biotic carbon storage is a stock-flow problem. The net change in carbon stock over one year can be viewed as the annual rate of carbon sequestration. Although a mature forest does not continue to absorb atmospheric carbon on a yearly basis, it does store a great amount of carbon. Newly planted forests tend to have high rates of carbon fixation, but decades if not centuries will pass before sufficient carbon is sequestered to replace that from the original destroyed forest.

Although global warming is by definition an international concern, we focus only on forestry efforts within the U.S. Forestation options abroad, such as halting *deforestation* in the tropics may be the most effective mitigation method both on a hectare-by-hectare and a cost per-hectare basis. Many reports mention the possibility of funding tropical reforestation efforts. However such programs are likely to be difficult to implement and monitor effectively. Moreover, attempting to model afforestation in developing countries is difficult for several reasons. The data needed may not be available. The domestic concept of property rights may make the results questionable. Lastly the development of an international model would necessarily increase model complexity. An inter-regional model of the U.S. likely would be computationally infeasible if other countries were also included.

Although this research does not explore international efforts, the exclusion is not meant to discredit the option. The creation of international afforestation projects can proceed independently from domestic tree planting efforts. Some of the major cost parameters, such as the land supply curve, should not be affected by forestation efforts abroad. Aside from interaction through trade of agricultural and timber products, the main link between domestic and foreign tree planting efforts would be competition for funding.

MERGE and other emissions abatement models have detailed energy sectors utilizing different technologies but do not offer the option of wealth transfers from one region to improve the energy efficiency of another. U.S. funding of foreign forestation efforts should be viewed in same light as other forms of joint implementation, such as providing monetary incentives to developing countries to build more energy-efficient plants or to develop less energy-intensive technologies and industries.

Section 2: Research Objectives

This research builds upon previous studies that attempt to calculate afforestation solutions in the form of top-down models. Over the past few years research on afforestation has evolved from its initial state of back-of-the-envelope calculations. Recent analyses incorporate at least some of the following features: rising marginal costs, long time horizons, discounting methodology, and issues of interaction between afforestation programs and the timber and agricultural markets. Richards et al (1993) attempt to model afforestation as part of an abatement program that includes traditional offset measures such as conservation and fuel switching. However none of the previous models examines all of these ideas and issues. Furthermore, the focus of the studies reviewed has been to determine feasible levels of afforestation. We address the question of what level of afforestation is *optimal*.

To analyze this question quantitatively we have constructed TREES, a model analyzing possible domestic afforestation projects. Given project costs, benefits from sequestration, and other parameters such as regional absorption rates, the model calculates the optimal afforestation project schedule. The policy recommendations are based on arguments of cost effectiveness. Through an analytical approach, this research effort aspires to contribute an important tool for evaluating the option of afforestation as a potential national mitigation strategy.

Integrated Assessment

Previous afforestation research has focused on reaching a fixed sequestration goal: either an annual sequestration rate for a set period of time (ignoring the fact that such programs are unlikely to have uniform sequestration rates over time) or a total sequestration goal to be reached by a specified time. However the use of fixed sequestration quotas may not be economically efficient. While policy makers may find it difficult to quantify the monetary benefits of carbon sequestration, examining the shadow price associated with the quota will reveal the price that is implicitly assumed.

The above approach can be improved upon. The use of annual net emissions goals is unnecessarily arbitrary. The need for emissions reduction in some years may be greater than in others. MERGE, a global integrated assessment (GIA) by Manne and Richels (1995), solves the problem by integrating a climate sub-model to evaluate the benefits of abatement. Afforestation measures are likely to supplement emissions abatement measures such as fuel switching. The international price of carbon is calculated by MERGE and represents the value of sequestering a ton of carbon.

Solving TREES with links to MERGE determines how much mitigation activity occurs in each time period and what percentage of abatement is derived from domestic sequestration efforts. The carbon tax of Richards, Moulton and Birdsey (1993) provides a link between the afforestation and traditional abatement activities. However this tax represents the shadow price associated with fulfilling an arbitrary, predetermined goal of stabilizing emissions. None of the previous studies utilizes a model in which the costs and benefits of fixing a unit of carbon are weighed against each other to determine if this unit should be sequestered or otherwise abated.

Time Horizon and Discounting

The time horizon of an afforestation analysis needs to be extended beyond the 30 to 50 years typical of many models. As shown in Richards, Moulton and Birdsey (1993), even some of the fastest growing trees have reached only half of their potential growth by then. Expansion constraints for tree planting will result in a staggered implementation of afforestation, further delaying the sequestration effects. TREES is designed to have a time horizon out to the year 2100 and possibly beyond.

Since the majority of the sequestration benefits from tree planting are reaped after program costs are incurred, appropriate discounting is needed for a cost-benefit analysis. This feature is especially important in models with long time horizons. The value of the discount rate can greatly affect model results. To be consistent with MERGE, we utilize a 5% discount rate. Some of the previous studies discount the costs of afforestation while measuring the benefits, the rate of sequestration, in undiscounted terms. We avoid this distorted view by discounting both the costs and benefits of afforestation.

Carbon Sequestration Time Path

The carbon uptake rate of growing trees varies over time and with tree species. The model utilizes regional, time-variant sequestration rates. These differing rates, local stumpage prices and land rental costs are the main inter-regional variances that motivate the regionalization of the model.

The increase in carbon storage is only part of the net flow of the carbon sequestration equation. The release of carbon from trees and wood products also needs to be examined. Most analyses either ignore trees that die or treat them as instantaneously decaying. Wastes from harvesting (e.g. roots, branches, and leaves) are often ignored as well. Harvested timber that is processed into finished products is often treated as permanently sequestered. Over a long time horizon these assumptions are unsatisfactory. Decay rates are incorporated in TREES. The carbon released in each time period can be tracked and used to adjust net sequestration.

Carbon prices are the informational link between MERGE, the more traditional emissions abatement model, and TREES. The level of these prices as well as their trajectory over time have a great effect on the optimal solution of the afforestation model. The time path of carbon sequestration (and release) will shift depending on many factors, especially whether the discounted social benefits from sequestration activities grow more slowly or quickly relative to their costs.

Interactions with Other Markets

One weakness of existing analyses is that interactions of afforestation projects with traditional timberland are not modeled. The harvest of trees from an afforestation project could influence the price of lumber and affect returns to the project and to the lumber industry in general. Decreased stumpage prices may in turn reduce forestation efforts. However the forgoing of all project harvests may not be the best solution. The government would pay project upkeep costs in perpetuity, and project planners would eventually run out of land for future plantings.

The goals of the timberland owners are not necessarily identical to those of the government. These individuals can be expected to maximize profits without consideration of the public goods problem of CO_2 emissions. Unless subsidies or other market interventions are used, the timber industry should be expected to act without reference to carbon abatement benefits. We attempt to capture the interactions between afforestation and existing forest planting and harvesting activities.

Afforestation projects also affect the agricultural sector. As mentioned by Adams et al (1993), large scale afforestation programs incur additional costs not represented by most studies. The costs result from increasing rental costs as afforestation programs compete with agriculture for the same land. TREES incorporates an additional cost curve based on total project land usage to address these concerns.

Section 3: Structural Overview

An overview of the model structure shows how we address the above goals. Illustrated at the conceptual block level in Figure 1, a reduced form model of the regional timber industry, *traditional timberland*, is linked with a model of domestic regional *afforestation*. The timber harvests from these modules are used in the *market* module, which calculates regional stumpage prices based on the demand function and the supply from both of the timber modules. The net carbon sink is calculated from the activities of the three modules in the *sequestration*

Figure 1: Block Representation of Modules and Objective Function for Each Region

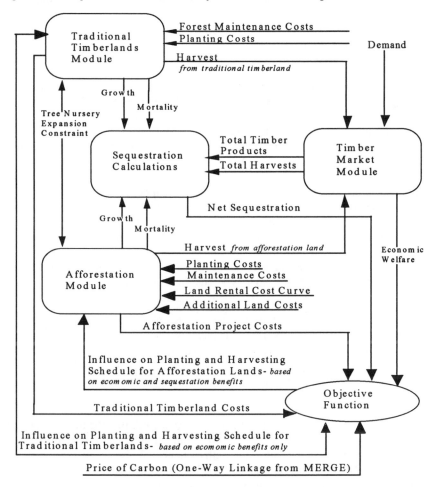

calculations. The two timber production modules are further linked by a nursery expansion constraint.

Solving the above structure produces a feasible solution. However an optimal solution is reached by maximizing the *objective function*. The objective function is measured with respect to two types of benefits, each in discounted 1990 dollars. Sequestration benefits are calculated through multiplying the net carbon offsets by the exogenously specified price of carbon. Economic benefits are measured as the sum of producers' and consumers' surpluses in the timber sector. The costs and revenues from both the timberland and afforestation lands are included in this measure of welfare. The solution is optimized through modification of the decision variables: the planting and harvesting schedules of both the traditional timberland and the afforestation project modules.

Traditional timberland module

This module represents the private sector's activities in timberland management. All existing timberland is modeled as property of the private sector. One of the complications with modeling forests as carbon sinks is the difference in the sequestration rates and carbon storage for different ages of timber stands. Young forests have accumulated little carbon but still actively absorb carbon from the atmosphere. Mature forests store a larger amount but no longer accumulate carbon. The model differentiates between stocks of different ages to describe the inherent stock-flow nature of this relationship. The variable $s_{t,a}$ measures the millions of hectares of timber stands of age **a** years at decade **t**. The model's seven age classes correspond to trees aged 0-9 years, 10-19 years, 20-29 years, 30-39 years, 40-49 years, 50-59 years or 60+ years. The last age group includes all trees over 60 years of age. These mature trees are virtually identical for the purposes of measuring carbon sequestration. Mature trees do not actively fix carbon. Life spans of trees are long with respect to the time horizon of the model. Ten year age classes are used to match the model's decadal time intervals.

In TREES stand projection, tracing the distribution of trees through time is based on mortality, harvests, first year plantings, natural growth and inventories from the previous period. The relationship is illustrated in block diagram form in Figure 2 and is comparable to a *cohort classification scheme*. Stand projection for the afforestation module is represented in the same manner, save that no natural regeneration is allowed.

To keep the model at a manageable level, we assume that the trees planted are the same species within a region. With this simplification, the model uses identical carbon fixation rates for both commercial forests and afforestation projects. We do not distinguish demand between these two sources of lumber. However based on the assumption that the traditional timber industry is likely more able to secure stumpage contracts, these producers are guaranteed at least 75% of the stumpage sales each period.

Afforestation Module

This module represents the public sector's attempts to grow trees for carbon sequestration. These efforts are separate from reforestation on traditional timberlands. Other possible afforestation policies (e.g. publicly financed stock improvement on private timber stands) are not examined in the model.

Figure 2: Balance of Stock Illustrated for Timberland and Afforestation Modules

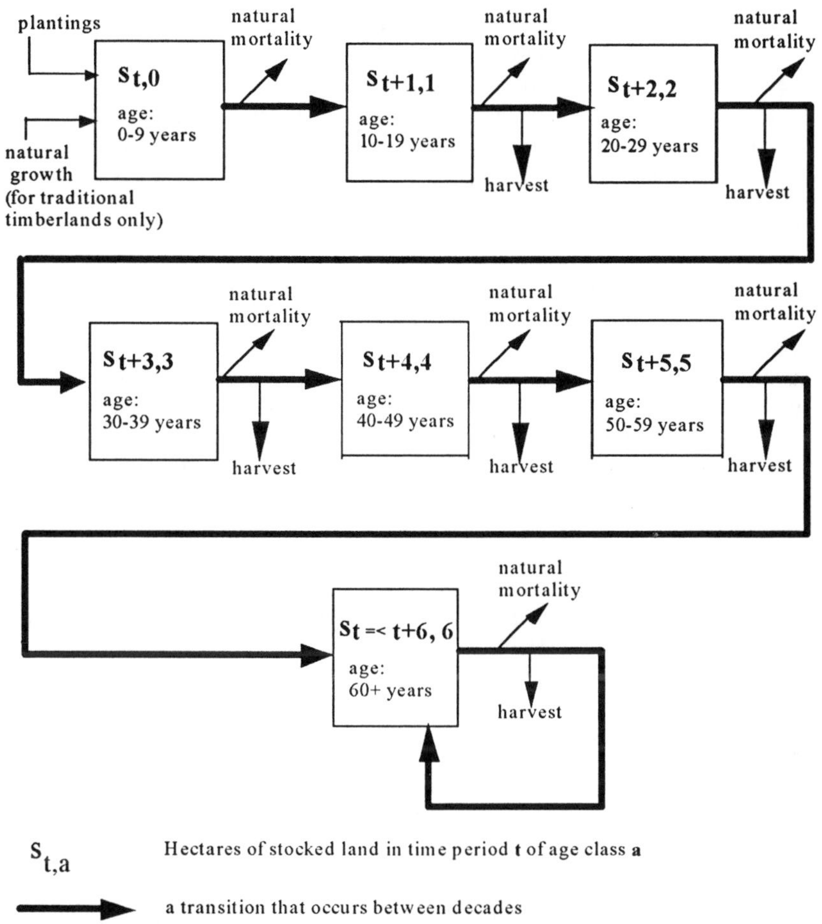

$S_{t,a}$ Hectares of stocked land in time period **t** of age class **a**

       a transition that occurs between decades

In addition to planting and maintenance costs, afforestation projects require land rental. Using of linearly increasing marginal costs for land rental captures the idea that afforestation costs rise with project scale as cheaper land supplies are depleted. Because afforestation projects require the removal of land from non-timber agricultural uses, competition for this land from farmers may increase land prices further. The marginal cost of land increases not only because the cheaper lands are utilized first, but also because the total supply of available land declines. This additional cost is based on research by Adams et al (1993). Traditional Timberland and Afforestation: Expansion Constraint.

In addition to interactions through the timber market, the afforestation and timber modules in TREES are jointly constrained by tree nursery capacities. Seedling production should not be expected to increase instantly to accommodate large-scale planting efforts. According to Moulton and Richards (1993) domestic tree planting efforts involved a record 1.4 million hectares in 1988 and strained nursery production capacities. Therefore the model has a dynamic constraint on the expansion of planting efforts. Nursery production within a decade is allowed to be twice that of the previous decade's. The model assumes that nurseries do not import or export seedlings to other regions. Each region is responsible for its own nursery production and expansion.

Timber Market Module

The market timber supply and price are determined endogenously in the *market module*. The domestic demand is represented by an inelastic CED (constant elasticity of demand) curve, normalized to the base supply and stumpage price predictions from the USDA (1990). At a given price the demand is not static but grows over time. The timber supply from the traditional timberlands is bounded below by the amount supplied from the scenario with zero carbon valuation. This lower bound prevents the timber industry from restricting supply because of consideration of the carbon benefits from decreasing harvesting.

Rather than calculating an explicit supply curve, TREES performs an intertemporal optimization. Producers maximize discounted economic benefits from supply, net of discounted costs. Traditional timber producers' costs are partly determined by the costs associated with replanting efforts and maintaining the standing inventories, as shown. Not explicitly shown in Figure 1 are the costs associated with the size of the traditional timber harvest each period. Higher rates of harvest result in rising costs. Per unit harvest costs decrease over time through technological improvements.

The price of timber will be higher than the sum of all the calculated marginal costs since the intertemporal optimization accounts for the opportunity costs of harvesting. Opportunity costs arise because an area harvested in decade **t** is not available for harvest in t+1 and the next few decades. Timber is renewable within the timeframe of the model. Therefore the opportunity costs rise over time but more slowly than the discount rate.

Not only is timber supply endogenous in TREES, but the age at which the timber is harvested is free to vary. Relationships from the market module as well as carbon valuation and other factors are used to determine the optimal harvest cycle. Harvest ages may change over time and differ between the timber and afforestation modules.

The cost structure for afforestation timber producers is similar to that of traditional timber producers. Additional expenses associated with project land usage are also included in afforestation costs. However the revenues from afforestation harvests are by-products of the primary purpose: increasing sequestration. Therefore an afforestation project may be larger than consideration of only the costs and revenues associated with stumpage production would dictate.

Sequestration Calculations

This module calculates carbon sequestration by examining the carbon flows from the traditional timber, afforestation, and market modules. The output from this module is not directly examined by the other modules. Instead, the relevant variable from the module, the net carbon sequestration over time, is used in the objective function. The net carbon sequestration in any period is modeled as the total sequestration from stand growth minus the carbon stock released to the atmosphere from decay of tree mortalities, harvesting wastes, and lumber products. The majority of studies consider only the first term, gross carbon fixation, in estimating the effectiveness of sequestration programs.

The model captures the stock-flow nature of carbon sequestration. Flows are represented by an age-variant carbon fixation rate to all of the timberland and afforestation project stock in the growth decades. The carbon stored (stock) in a stand is represented by the sum of the carbon fixation rates (flows) from the periods that the stand has grown.

Carbon stock may be released back into the atmosphere as the wood decays. This release is non-instantaneous and may depend on whether carbon is from trees that have died of natural causes, harvesting waste or timber produces. The model captures the gradual nature of decay and the fact that some of the carbon stock, e.g. long lived timber products, may effectively be sequestered permanently.

Objective Function

With the two potential timber producing sectors making decisions according to different objectives, we attempt to model the different behaviors resulting from this segmentation of the timber market. Both sequestration and economic benefits are considered for the decisions in the afforestation projects module. We assume that the private sector, unlike the public sector, does not place value on carbon sequestration. While carbon sequestered in the private timberlands has the same benefit per ton as that sequestered in publicly-funded afforestation projects, the private timberland owners will presumably not receive afforestation funding from the public. Therefore the private timber growers make planting decisions without considering the value of carbon.

Unlike other afforestation models, TREES employs a cost-benefit approach instead of one that seeks to solve for a predetermined sequestration goal. Through the time-variant parameter $CVAL_t$, the unit value of carbon abated, the model has a one-way link to a global integrated assessment (GIA) model, such as MERGE by Manne and Richels (1995). This link represents one further step towards a fully integrated assessment of afforestation with climate impacts and other mitigation measures.

If the carbon sequestered in afforestation projects is sufficiently small so as not to affect carbon prices, then only a one-way linkage with a global integrated assessment model is required. Given that the carbon prices in MERGE are assessed on a global basis and that model results show domestic afforestation efforts are a small part of a mitigation strategy, explicit coupling between the models is unnecessary. Should the amount of carbon sequestration from afforestation be sufficient to affect world carbon prices, then feedback to MERGE or another GIA model would be needed.

Section 4: Regionalization and Model Data

While many studies analyze the United States as a single region, TREES divides the nation into the four regions of Figure 3. These regions, the North, the South, the Rockies and the Pacific, are based on the divisions used by the USDA (1992) and Moulton and Richards (1990). Differing regional sequestration rates and cost characteristics, shown in part in Table 1, define a generalized character of the region and its suitability for afforestation projects. The regional sub-models are solved in parallel; each region is solved independently without feedback from the other three regions. Regional projections are then aggregated into national totals.

Figure 3: Model Regionalization of the United States

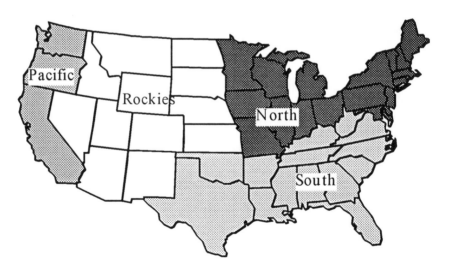

The obvious reason to regionalize a model is to obtain a more detailed solution. However an additional result is that that some of the costs and constraints that are inherently regional have a greater effect on the multi-region model's solution. Inherently regional constraints or costs are more likely to restrict multi-region models because some regions, such as the South, may be responsible for most of the sequestration efforts. By ignoring inherently regional effects, a single region model may produce a solution that underestimates costs and overestimates project feasibility.

Adding regionality within a single national model increases the dimensionality and adversely affects computability. Parallel regional models are solved in much less time than a multi-regional national model. This shortcut has the limitation that interactions between the regions are not captured. In TREES, however, afforestation efforts are based on a net benefit approach rather than on satisfying an exogenously specified goal. This approach would still be an oversimplification if U.S. sequestration efforts were sufficiently large to affect the valuation of carbon. Such is not the case, and changes in sequestration activities between regions of the U.S. would have little effect on carbon valuation. Therefore planting levels in one region can be represented as independent of efforts in other regions. Likewise the timber and agricultural sectors are modeled as having no inter-regional dependencies. The data used in TREES support this conjecture, as shown in Swinehart (1996).

Table 1: Regional Differences in Parameters

	South	Pacific	North	Rockies
Representative Species	Loblolly pine	Douglas fir	Blue spruce/ White pine	Ponderosa/ Larch pine
Timberland (M Ha)	80.7	28.14	63.9	25.4
1990 planting rates (M Ha/yr)	0.81	0.22	0.057	0.057
Percent share of national market	50%	30%	11%	9%
1990 stumpage values ($/tC)	113.84	102.79	27.63	34.26
Land available for afforestation (M Ha)	35.0	4.1	52.8	15.8
Project establishment costs ($/Ha)	162.13	450.00	335.58	218.47
Range of rental costs ($/Ha*yr)	31.44 to 150.67	120.00 to 143.00	51.38 to 200.70	32.38 to 128.44
Carbon sequestered after 60 years growth (tC/Ha)	150.3	373.0	125.2	94.7

Southern Submodel Data

The South, with less than a quarter of the national land area, contains 40% of U.S. timberland. Southern timberland is generally more intensely managed than of the other regions, with over 70% of all tree planting efforts occurring here. The South has a 50% share of the national timber market, and 1990 per-volume stumpage values are the highest. Southern per-hectare sequestration potential is the second highest nationally. Additionally Southern growth rates, typified by that of the Loblolly pine, are the fastest in the first few decades. With the lowest project establishment costs and low land rental costs, the South appears to be a good candidate for afforestation. Tree plantations are currently being established in the

South without consideration of sequestration benefits. The South has 40% of the 108 million hectares of marginal agricultural land that Moulton and Richards (1990) consider for afforestation.

Pacific Submodel Data

The Pacific region is the smallest in size, at 9% of the national land area. However this region is the next largest supplier of timber, producing 30% of domestic supply and having a per-volume stumpage value close to that Southern stumpage. Only 15% of U.S. timberland is in this region, but per-hectare productivity is higher than in any other region. About 20% of all domestic tree planting occurs in this region. Little land is available for afforestation. Land is more expensive than in the South and establishment costs are the highest in the nation.

Northern Submodel Data

The North has the second largest amount of timberland. Over 30% of U.S. timberland is located in this region, which accounts for 18% of U.S. land area. Unlike the South and Pacific, the North is not a major timber supplier, and produces only 11% of domestic timber on the market. Northern timber has the lowest stumpage value in the nation in 1990. Therefore little effort goes into replanting, with only 5% of national replanting activity occurring in the North. Although the North contains almost half of all the marginal agricultural land listed by Moulton and Richards (1990), the region is not as likely a candidate for afforestation. Land rental and establishment costs are high, and per-hectare sequestration potential lags behind both the Pacific and the South.

Rockies' Submodel Data

This region also includes the Great Plains and is the largest region examined, comprising 32% of total U.S. land area. Despite this size, the region has the smallest amount of timberland, only 13% of the national total. This region is the smallest timber producer, with only 9% of the national timber supplied to market. Stumpage values are low, partially the result of the high costs associated with harvesting stands in remote, mountainous areas. Not surprisingly, replanting efforts are minimal. About 15% of the nationally available project land is located in this region. Rental costs are comparable with those in the South, with establishment costs somewhat higher. Per-hectare productivity (and sequestration potential) is the lowest in the nation.

Linkage to MERGE: Carbon Valuation

The carbon sequestration benefits are computed from $\mathbf{CVAL_t}$, which is specified exogenously. The values represent the time path of the optimal carbon tax, given the costs of CO_2 abatement and the benefits of mitigating temperature increases. Since this valuation is uncertain and model results hinge upon it, we analyse scenarios with differing carbon trajectories. Two different carbon trajectories are imported from MERGE. Per-period U.S. emissions and abatement levels are also imported and are illustrated in Figure 10 and Figure 11. The nominal carbon valuations, with MERGE's time periods interpolated for decadal values, are shown in Table 2.

The baseline scenario utilizes carbon valuations from the Pareto optimal baseline case in MERGE with a discount rate of 5%. With this scenario a moderate amount of abatement occurs, especially after 2040, when non-electric carbon free energy sources are projected to be available. Carbon values are moderate, starting at $6/tC (tons Carbon) and rising to under $100/tC by the end of the 21st century.

Table 2: Carbon Valuation ($/ton C) for MERGE Scenarios

year	Base Valuation Scenario	High Valuation Scenario
2000	6.25	114.34
2010	8.50	160.52
2020	11.20	217.98
2030	14.56	292.36
2040	18.26	376.98
2050	21.60	466.26
2060	31.70	684.55
2070	40.73	902.84
2080	54.60	1228.71
2090	72.78	1662.16
2100	90.96	2095.62

The high valuation trajectory is from the scenario in MERGE that assumes climate sensitivity to emissions is high and that economic damages from global warming are severe. Although damages occur mainly after 2050 and a discount rate 5% is used, potential damages are sufficiently severe to prompt immediate reductions in emissions. Emissions are nearly eliminated by 2050, when an expensive but carbon-free energy source is assumed to be available. Carbon valuation increases by an order of magnitude over the baseline scenario, starting at $114/tC and rising over $2000/tC by 2100. For further detail on the assumptions used to generate the carbon values, refer to Manne and Richels (1995).

Section 5: Results and Conclusions

This section analyzes the results from both the baseline and high carbon valuations from MERGE. Sequestration rates are compared with emissions and abatement rates. Table 3 provides a summary of the results and a side by side comparison for the two scenarios. A comparison of results from TREES and MERGE provides the opportunity to define the role of domestic sequestration in a national mitigation policy.

Measuring Program Effectiveness: Net Sequestration

TREES tracks carbon sequestration by U.S. commercial timberlands as well as interactions between commercial timberland and afforestation projects. Traditional timberland harvests are not based on sequestration benefits. Therefore lower bounds for these harvests are set by running the model with a carbon valuation of zero and without afforestation planting allowed. Without this lower bound, the model would erroneously reduce traditional timber harvests in scenarios with positive carbon valuation.

Table 3: Summary and Comparison of Base and High Valuation Scenarios

	Base valuation scenario	High Valuation Scenario
Range of $CVAL_t$	6.27 to 90.61 $/tC	114.34 to 2095.62 $/tC
Maximum land usage	20.3 million Ha	103.0 million Ha
Total carbon offset	0.321 billion tC	11.826 billion tC
Discounted carbon offset	0.009 billion 1990 tC	0.721 billion 1990 tC
Total undiscounted value of carbon benefits	11.10 billion $	9811.7 billion $
Total discounted value of carbon benefits	0.304 billion 1990$.	295.32 billion 1990$
Total discounted cost	1.591 billion 1990$	74.35 billion 1990$
Total discounted timber revenue from afforestation	1.256 billion 1990$	0.42 billion 1990$
Percent of total project benefit from timber revenue	80.5%	0.1%
Levelized per ton cost	36 1990$/1990tC	103 1990$/1990tC
Total discounted change in welfare over benchmark	0.004 billion 1990$ (net gain)	-74.60 billion 1990$ (net loss)
Improvement over benchmark scenario	0.308 billion 1990$	220.72 billion 1990$

Gauging the effectiveness of an afforestation program (in terms of carbon sequestered and net benefits) requires comparison against a benchmark. Scenarios are compared to a "business-as-usual" run, with afforestation planting fixed at zero, but with all other parameters unchanged. The difference in carbon sequestration is attributed to the program's effectiveness. Since the business-as-usual case has an additional constraint, the value of the objective function will be lower than in the unconstrained case. This approach is a simplification because tree planting on land designated for afforestation may occur regardless of whether or not sequestration projects are funded. To determine the effectiveness of the carbon valuation benchmark, a scenario is examined with a zero carbon valuation and with afforestation allowed. Results show that effectively no additional carbon is sequestered under this scenario. Therefore the metric used in this chapter for measuring program sequestration is satisfactory.

Results for the Scenario with Baseline Carbon Valuation

The following scenario uses a carbon valuation, $CVAL_t$, derived from the carbon tax specified by the pareto-optimal case of MERGE. In this case, Manne and Richels (1995) set to a moderate level both the climate sensitivity to increased atmospheric carbon and the economic damages from climate change. Hereafter this scenario is termed 'baseline' and is deemed the most likely of the ones explored. The nominal value of carbon rises over time but decreases in present value terms at a 5% discount rate.

With this carbon valuation the optimal afforestation program is of modest scope when compared with many other studies. Figure 4 shows that annual afforestation efforts are never above 1.2 MHa/yr (million Hectares per year). These efforts are a small fraction of total planting activities. Afforestation planting takes at most 25% of seedling production even in the peak years of project establishment. These moderate efforts are not constrained by nursery expansion limitations. New planting efforts are constrained only in the Rockies and the North, two regions where little or no sequestration occurs. Most afforestation planting occurs in the South. Afforestation planting is highest in the years 2060 to 2080. The planting rate drops significantly after 2080 because trees planted thereafter are not economical to harvest by 2100.

Figure 4: Annual Afforestation Planting Rates for Base Scenario

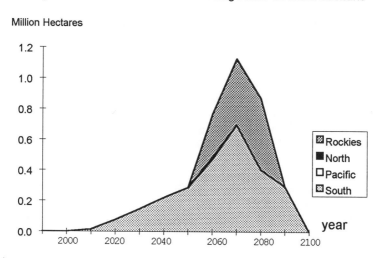

Figure 5 shows that total usage of the afforestation land never exceeds 21 million hectares. This usage is well below the national capacity of 108 million hectares. Land constraints are not active even in the regions with significant afforestation. Although some harvests of project trees occur before 2060 (mainly in the South from early harvest of its rapid growth trees) afforestation land usage slowly increases until 2080. The afforestation land is harvested faster than it is replanted, decreasing project land utilization thereafter. By the end of the model's time horizon, just over five percent of the potential project land is forested. The low remaining inventory is partially the result of the terminal condition imposed in the model: discounted rent payments continue indefinitely on project land still forested in 2100. Afforestation efforts would increase slightly with this constraint removed.

Figure 5: Afforestation Land Usage for Base Scenario

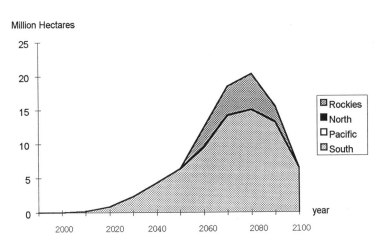

Examining land utilization by region shows that afforestation has distinct regional characteristics. No afforestation occurs in the North, which has high rental costs and low stumpage values. A small amount of the available land is utilized in the Pacific (less that 0.2 million hectares, 5% of available land). Although most of the planting efforts and inventory occur in the South, at most 15 million hectares, under half of the available land, are forested at any time. Less than a third of project land is planted in the Rockies, and this activity occurs only in the last decades of the 21st century.

Afforestation harvest age varies between regions. Afforestation lands are harvested before maturity in the South. In fact harvests on Southern project land occur before the present value of stumpage is maximized. A comparison of the discounted benefits from sequestration with the discounted stumpage revenues reveals that 80.5% of the total benefits are from stumpage revenues. Additionally the land rental costs encourage earlier harvests of project land. Although afforestation projects are being proposed to absorb carbon, the side benefit of stumpage revenues is of greater value. The optimal afforestation program will be focused more on maximizing revenue potential than on maximizing sequestration.

One of the original premises of this research was that harvests from afforestation projects might flood the timber market. Therefore we include a constraint that guarantees traditional timberland a 75% share of all regional stumpage markets. However this constraint is not binding in the South, which has the largest afforestation projects for the baseline scenario. The constraint is binding in the Rockies from 2080 through 2100, but this region has the smallest timber market. The afforestation projects in the Pacific contribute little to the timber market. Therefore that regional constraint is not binding.

Since the afforestation supply constraint is not binding in most cases, the decision to limit project size below capacity is the result of rising marginal project costs rather than market limitations and interactions with traditional timberlands. Given that the project size is well below capacity, it is unlikely that the Adams's costs (the additional costs from land usage) play a significant role in limiting project size. Marginal annual rent rises as high as $83 per hectare, while the marginal contribution from the Adams's costs is only $21 per hectare.

To measure the effective carbon offset from the afforestation program, the baseline scenario is compared to a run that disallows planting on potential project lands. Over the 110-year time horizon, afforestation is responsible for sequestering a total of 0.3 BtC (billion tons Carbon). This carbon offset accrues over time and by region as shown in Figure 6. Yearly sequestration rates rise over time, peak at 2090, and then sharply decline. Nearly all (96%) of the sequestration occurs in the South. While afforestation planting is next highest in the Rockies, those stands are harvested early. Little net sequestration occurs in that region or in the Pacific.

Figure 6: Annual Carbon Sequestration Rates for Base Scenario

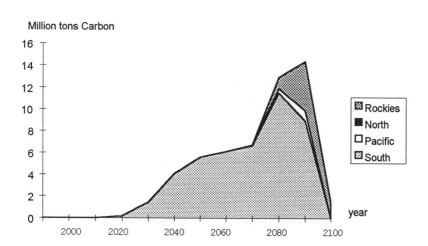

As shown in Table 3 the net sequestration benefits from afforestation projects total just over 11 billion dollars. However most of these benefits are realized in later years. A more meaningful metric is the cumulative discounted sequestration benefit over time. These present value benefits tally 0.304 billion 1990 dollars. The other benefit measure in the objective function is economic welfare: the area under the demand curve minus production costs from both the traditional timber industry and afforestation projects. The present value of welfare improvements is 0.004 billion 1990 dollars.

The discounted revenue from afforestation is 1.256 billion 1990 dollars. Added to discounted sequestration benefit, this amount represents the discounted benefit from the implementation of afforestation projects. Therefore in this scenario most (80.5%) of the total benefits from an afforestation project are from timber revenues, not carbon sequestration. The discounted cost of afforestation is 1.59 billion 1990 dollars. We subtract the discounted revenue from discounted cost and divide by discounted carbon sequestered (9.3 million tons discounted to 1990) to calculate a levelized cost similar to that used by Stavins (1995). This is $36/ton for the baseline scenario of TREES. This metric fails to capture the dynamic nature of carbon valuation and sequestration rates but is useful as a simple comparison statistic.

Results for the Scenario with High Carbon Valuation

In contrast to the baseline scenario, this scenario utilizes the carbon tax specified by the high damage case of MERGE. Carbon sequestration has a higher value in this scenario, resulting in afforestation projects on a much grander scale. Every region establishes projects, and planting would occur sooner in some regions if not restricted by the nursery expansion constraints. These constraints are responsible for the two "spikes" that appear in the annual project planting rates as shown in Figure 7. The first major planting effort occurs in 2010 to 2020 in the South and the Pacific because expansion constraints are still active in the Rockies and the North. The second major planting effort occurs between 2040 to 2050, when these regional constraints are no longer binding. Afforestation planting efforts are greatly reduced after 2060 since they only replace trees harvested or lost to mortality.

The planting efforts result in a huge expansion of nursery production. In the peak years of project establishment, afforestation plantings account for 33% to 62% of all planting efforts. Annual southern afforestation planting efforts in 2010, 2.1 million Ha/yr, exceed the traditional timberland replanting rate, 1.04 million Ha/yr. Northern annual afforestation efforts in 2040 and 2050, 1.7 and 2.01 million Ha/yr respectively, dwarf the traditional timberland replanting rates at these times, at 0.12 and 0.25 million Ha/yr respectively. (Northern replanting rates are limited by low stumpage values, especially in the early years.) The high afforestation planting rates differ greatly from those of the base carbon valuation scenario, where afforestation efforts encompass at most 25% of total nursery production at any time.

Figure 7: Annual Afforestation Planting Rates for High Scenario

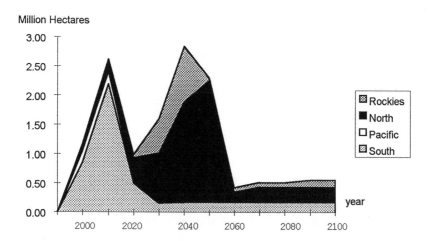

These active planting efforts are reflected the utilization of project land over time, shown in Figure 8. Afforestation increases until 2050, when effectively all of the available land is planted. The South and the Pacific achieve full capacity by 2020 and 2010, respectively, while the Rockies and North take until 2040 and 2050 to use all available land.

Figure 8: Afforestation Land Usage for Base Scenario

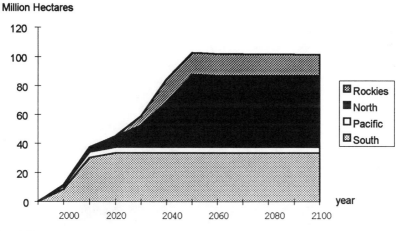

The full utilization of the project land results in high Adams's costs. The competition for the remaining project land drives marginal costs up by 60% and more. (Total land costs are between 30% and 50% higher on account of the Adams's effect). However the carbon benefits received outweigh these additional costs. Likewise the inclusion of the terminal rent condition for afforestation has no effect on the decision variables. Afforestation is constrained by land availability, not project costs, in this scenario.

Unlike in the baseline scenario, harvests occur on afforestation land only in the Rockies. In that region harvests occur on less than 10% of project inventory and only after the trees have reached maturity. Stumpage revenues account for only a negligible percentage (0.1%) of total project benefits. This is in stark contrast to the high percentage (80.5%) of benefits from afforestation revenues in the baseline scenario.

The gain in net total carbon sequestration from the optimal afforestation scenario as compared with the business-as-usual case over the 110-year timeframe is 11.8 BtC. Figure 9 shows the annual rate of carbon sequestration over time. The annual rate rises quickly until 2020, rising more slowly until 2070 and declining thereafter. In the peak years annual sequestration rates range between 130 MtC and 170 MtC. For perspective the peak absorption rate is one-tenth that of annual U.S. energy emissions of carbon in 1990.

On a regional basis, the South sequesters a total of 4.4 BtC over the model's timeframe. But the North makes a larger, although delayed, contribution totaling 4.9 BtC. Figure 9 provides an understanding of the regional distribution of carbon benefits measured in present value terms. Present value carbon valuation decreases over the timeframe. Therefore the South's contribution will have greater value than that from the North. The Rockies and the Pacific contribute approximately the same amount, 1.2 BtC each. Although the Pacific has little project land (4 million hectares), this region has a higher per-hectare offset potential. The offset from any one of these regions dwarfs the national carbon absorption in the base valuation scenario.

Figure 9: Annual Carbon Sequestration Rates for Base Scenario

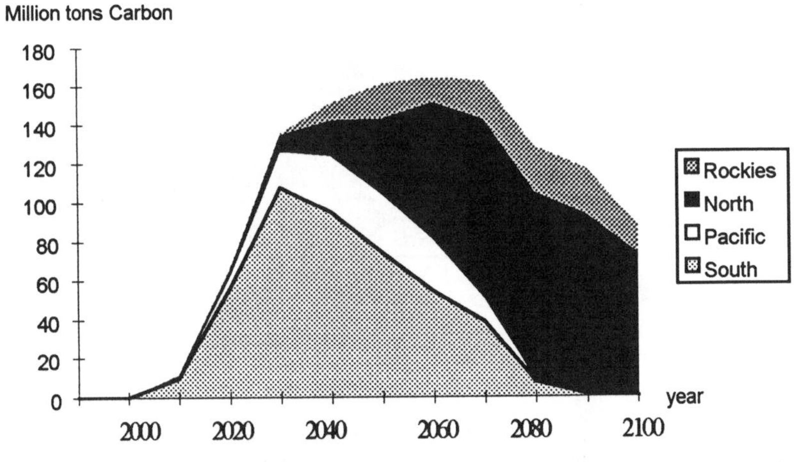

Million tons Carbon

Little or no afforestation timber is harvested in this scenario. There is little interaction between afforestation projects and the traditional timber industry other than the competition for nursery production. (The constraint on seedling production delays implementation of Northern afforestation projects.) The traditional timberland planting schedule is similar to that from the scenario disallowing afforestation. Not all vacant traditional timberland is replanted in the

North, since stumpage values do not yield sufficient discounted returns to merit investment. Replanting of Northern timberland is limited for all scenarios, regardless of the carbon valuation or if afforestation occurs. It should be noted that the shadow price is very high on the constraint that keeps traditional timberland harvests rates above those for the zero carbon valuation case.

From Table 3, recall that undiscounted carbon benefits total over 9800 billion dollars over the business-as-usual scenario. A better measure of sequestration benefits is the net present benefit from afforestation. This is 295 billion 1990 dollars when discounted at 5%. Countering these benefits are the large costs of the projects, discounted to a total of 74.6 billion 1990 dollars. With a discounted total carbon offset of 0.72 billion 1990 tC, the levelized cost per ton of carbon is $103. This is significantly higher than the $36 of the baseline scenario. The high valuation scenario results in a net loss in economic welfare, which almost exactly mirrors the project cost. This is not surprising given that there is little change in either harvest levels or stumpage prices. Unlike the baseline scenario, nearly all of the benefits of afforestation are the result of carbon sequestration.

Sequestration as a National Mitigation Strategy: the Big Picture

Previous studies discuss using sequestration either to offset emissions increases over 1990 levels or to absorb a significant fraction of current emissions. In the only research reviewed that explicitly incorporates other abatement strategies, Richards et al (1993) find that afforestation accounts for over half the net carbon reduction. Additional data from MERGE establish how the role of optimal afforestation efforts modeled by TREES compares with that of other abatement activities. Figure 10 shows the U.S. energy emissions rates for three scenarios: a business-as-usual case without abatement and the optimal emissions paths for the two scenarios with different carbon valuations. It should also be noted that emission levels are measured only for energy-related emissions. The projected decrease in timberland stocks (which leads to net deforestation) is not included in these figures.

Subtracting the emissions from a scenario with abatement from the business-as-usual emissions yields the rate of abatement. The offset from domestic afforestation can then be compared with these abatement levels. The MERGE scenario that produced the low carbon valuation has a total cumulative abatement from 1990 to 2100 of 44 BtC. Sequestration is responsible for at most 2% of the net abatement activities in any decade of the time frame. Figure 11 provides a graphical view of sequestration efforts compared with abatement. Since we cannot represent zero values on a semi-log scale, periods with no abatement or sequestration activity are not graphed

Figure 10: Three U.S. Emissions Scenarios from MERGE

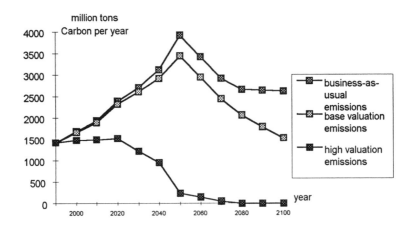

Figure 11: Two Scenarios for Abatement and Sequestration

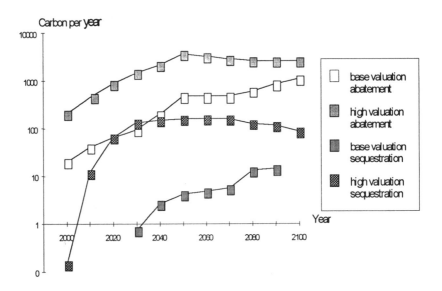

The MERGE scenario from which the high carbon valuation is derived has a total abatement over 110 years of 229 BtC. Afforestation efforts for this scenario

are able to sequester only 5% of this amount. However from 2060 onward, emissions in MERGE are zero. Therefore sequestration is the only way to reduce cumulative net emissions further. We can compare the sequestration rates from Figure 11 to the high valuation emissions path in Figure 10. From 2050 onward, afforestation and abatement in combination reduce U.S. net emission levels to zero or below for the high valuation scenario.

Although the size and nature of afforestation efforts from the baseline scenario differ greatly from those of the high valuation scenario, an important result holds for both. Afforestation is not likely (and, in the case of the high valuation scenario, not able) to play a large role in total national mitigation efforts. This finding contradicts some of the earlier studies, such as Richards, Moulton and Birdsey (1993). This limited role, along with the fact that carbon prices are likely to be determined through international joint implementation (e.g. *global* trade of emissions rights), justifies the use of carbon valuations that are independent of sequestration efforts.

Section 5: Conclusion

The purpose of a model like TREES is to obtain insights into the optimal afforestation policy. The levels of the decision variables rather than the absolute value of the objective function are the most important results for policy makers. Initial conditions, such as the age distribution of the existing forest lands, may have dramatic effects on the objective value, but changes in these parameters have little affect on the decision variables. However other parameters, such as the carbon valuation trajectory, greatly influence the optimal schedule.

As shown, the model is highly sensitive to the valuation of carbon. Both the magnitude and time path of carbon valuation influence the optimal afforestation planting and harvesting schedule. In the base valuation scenario the size of the afforestation program is determined not by feasibility but by the optimization of net benefits. In the high valuation scenario afforestation is constrained by land limitations. Feasibility becomes the limiting factor. For both carbon valuations considered, the net discounted benefits of the optimal afforestation program outweigh those of the optimal scenario without afforestation. However, this research shows that while afforestation may play a role in mitigation strategy this role will be limited.

References

Adams, R.M., D.M. Adams, C.C. Chang, B.A. McCarl, and J.M. Callaway, "Sequestering Carbon on Agricultural Land: A Preliminary Analysis of Social Cost and Impacts on Timber Markets," *Contemporary Policy Issues*, Vol. XI, No. 1, January 1993.

Global Environmental Change Report, Cutter Information Corp., Vol. VII, No. 22, November 23, 1995.

Manne, Alan, and Richard Richels, "The Greenhouse Debate- Economic Efficiency, Burden Sharing, and Hedging Strategies," *The Energy Journal*, 1995.

Moulton, Robert J., and Kenneth R. Richards, "Costs of Sequestering Carbon Through Tree Planting and Forest Management in the United States," General Technical Report WO-58, United States Department of Agriculture Forest Service, Washington D.C., 1990

Moulton, Robert J., and Kenneth R. Richards, "Accelerated Tree Planting on Non-industrial Private Lands," *White House Conference on Global Change*, Washington D.C., July 1993.

Richards, Kenneth R., Robert J. Moulton, and Richard A. Birdsey, "Cost of Creating Carbon Sinks in the U.S.", *Energy Convers. Mgmt,* Vol. 34, No. 9-11: pp 905-912, 1993.

Richards, Kenneth R., Donald H. Rosenthal, Jae A. Edmonds, and Marshall Wise, "The Carbon Dioxide Emissions Game: Playing the Net," May 26, 1993.

Stavins, Robert N., "The Costs of Carbon Sequestration: A Revealed-Preference Approach," [draft] September 1995.
Swinehart, Susan., "Afforestation as a Method of Carbon Sequestration: A Cost-Benefit Analysis," doctoral dissertation, Department of Operations Research, July 1996.

USDA Forest Service, *An Analysis of the Timber Situation in the US 1989-2040: A Technical Document Supporting the 1989 USDA Forest Service RPA Assessment,* Rocky Mountain Forest and Range Experiment Station, Fort Collins, CO., General Technical Report RM-199, 1990.

USDA Forest Service, *Forest Resources of the United States, 1992* Rocky Mountain Forest and Range Experiment Station, Fort Collins, CO., General Technical Report RM-234, 1992.

Chapter 8

ECONOMIC INCENTIVES, INTERGENERATIONAL ALTRUISM AND SUSTAINABILITY

Gunter Stephan and Georg Müller-Fürstenberger[1]
University of Berne, Germany

Abstract: *Within the framework of a small scale numerical model this paper carries out a simple thought experiment. There are two perspectives: one, in which investment-saving decisions are made by an immortal agent acting as trustee on the behalf of both present and future generations; and one in which generations simply save during working years and dissave during retirement, but where carbon rights are traded internationally on competitive markets. Despite these polar views on intergenerational altruism virtually the same results can be observed with respect to sustainability as well as the variables directly relevant to the greenhouse problem.*

1. Introduction

In his paper "Equity, Efficiency and Discounting" Alan Manne (1996) distinguishes between two different approaches to the economics of global change. In his view a descriptive approach places the global climate problem and its solution into the framework of a decentralized market economy. Economically efficient abatement policies are determined through cost-benefit considerations. No major changes are proposed in the ownership of labor, capital or other resources, and the market rate of interest is used for evaluating both conventional and environmental capital formation.

A prescriptive approach relates the greenhouse issue to the ideas of equity and intergenerational fairness. Typically, a prescriptive model assumes that the society can mandate wealth transfers between and across generations, and addresses the analysis to questions such as: How should the costs of greenhouse gas abatement be shared between different nations and different generations? What might be an equitable arrangement for cost sharing?

Sustainability itself is an ethical, or to put it into Alan Manne's classification, a prescriptive view on the classical problem of intertemporal optimization. In its original form, sustainability requires that the welfare of the present generation should be maximized without a negative impact upon the future generations' ability to meet their own needs (see World Commission on Environment and Development 1987).

The scientific community has not yet achieved full agreement, how to transform the above Brundtland Report's notion of sustainability into economically operational terms. But there is general agreement that sustainability must be viewed as a matter of intergenerational fairness. Generations should be treated symmetrically and equally. Constant per capita utility is therefore a minimal requirement. Moreover, if utility were a continuous and non-decreasing function of consumption, then sustainability would be equivalent to constant or increasing per capita consumption (see Den Butter and Hofkes 1995). This is just the view on sustainability we will adopt in this paper.

Independent of the concrete formulation, however, unless each generation is committed to transfer sufficient environmental and/or physical capital to the next, development cannot be sustainable. From an economist's point of view it is unreasonable to believe that the necessary intergenerational transfers can be based solely on the individuals' behavioral attitudes such as environmental ethics or intergenerational altruism. Global climate is a public good and individuals cannot provide for the living conditions of their offspring by acting independently and privately. Instead, arrangements for abatement and cost sharing between generations are required (see Howarth and Norgaard 1992).

New international institutions might be required for such an agreement. One reasonable option is suggested in Manne (1996, p.3): "Suppose that it were feasible to separate the issue of efficiency from that of equity. The time path of global emissions could then be determined independently of the contentious issue of how to ensure fairness in abatement cost sharing. An international technical organization might recommend the number of carbon emission permits to be distributed on an annual or five-year basis. These permits would in turn be tradable on an open international market. The worldwide rate of permit sales could be based on principles of intertemporal efficiency - that is through a comparison of discounted costs and benefits."

This paper carries out a simple thought experiment. A descriptive approach is contrasted with a prescriptive view on the issue of sustainability. On the one hand, a market -oriented approach to the solution of global environmental problems is considered. Internationally tradable carbon rights are defined on an annual basis such that global carbon concentrations are stabilized by the end of the 21st century. Generations have no altruism, but simply save during working years and dissave during their retirement. On the other hand, a scenario with complete intergenerational altruism is considered. The intertemporal distribution of natural resources and environmental capital is determined as if generations have delegated their investment and savings decisions to an immortal agent who acts as a trustee on behalf of both present and future generations.

2. Sustainability, Discounting and Selfish Egoism

Given these polar views on intergenerational altruism, three questions suggest themselves. (1) Is intergenerational altruism a necessary condition for sustainable development? (2) Do low social discount rates automatically guarantee ecological sustainability in the sense that global carbon dioxide concentrations never exceed doubling the pre-industrial level? (3) How can the different views on intergenerational altruism be captured by a theoretical framework?

There are different ways to focus on intertemporal decision making, intergenerational equity and efficiency. One option would be to employ an overlapping generations structure. A second possibility is to use a Ramsey approach and to start from the assumption that future generations are represented by a single infinitely-lived agent.

With respect to intergenerational altruism, the infinitely-lived agent (ILA) and the overlapping generations (OLG) approach represent opposite views. With ILA, an immortal representative agent takes investment/savings decisions as a trustee on behalf of future generations. In this way, the issue of intergenerational equity is automatically incorporated. Equity between generations is represented simply by adding up the generations' utility levels, where each generation's utility depends only on its own consumption, and the discounted sum of instantaneous utilities is taken as a measure of social performance (see Solow 1986). In other words, the social discount rate expresses the degree of intergenerational altruism. The lower it is, the more do present generations care for the welfare of future generations.

There are at least three objections to ILA. First, it is difficult to believe in the fiction of an immortal agent who views the increments in any future generation's utility as though they were increments in his own (see Schelling 1995). Second, as Howarth and Norgaard (1992) argue, a Ramsey-type cost-benefit analysis is inadequate, since it is based on the assumption that the discount rate can be selected exogenously. The discount rate is, however, determined endogenously. It depends on the degree of intergenerational altruism as well as on the productivity of capital and on the intergenerational distribution of assets. Finally, an ILA approach is inappropriate in analyzing those externalities arising from consumer heterogenity and disconnectedness across generations.

OLG avoids these problems. In particular, it sidesteps the issue of discounting intertemporal consumption or that of selecting a social discount rate. Each age cohort maximizes its individual life-time utilities. Individuals save during their working years, dissave during retirement and need not behave altruistically.

Even if agents put no weight on the distribution of intergenerational welfare, development can be sustainable over the long run. Independent of the specific numerical parameters, our computational experiments with OLG show that atmospheric CO_2 concentrations are eventually stabilized, and per capita consumption grows at the rate of productivity improvement. All that is required is the recognition that abatement represents a specific form of environmental capital accumulation, and that environmental regulations are implemented in an appropriate, market-oriented way through tradable quota rights.

In this paper twice the pre-industrial level of atmospheric CO_2-concentration is considered as the maximum 'ecologically sustainable' level. This definition is motivated by climatologists' claim that interglacial interactions are likely to become disastrous, if the atmospheric concentrations significantly exceed 550 ppm (personal communication with F. Joos, Dept. of Climate and Environmental Physics, University of Berne). Our OLG stabilization scenario is close to this threshold and therefore avoids high uncertainties in climate damage estimates.

Surprisingly, development need not be ecological sustainable in the just mentioned sense, even if agents behave altruistically. Our calculations suggest that this level is attainable only if the social discount rates are close to zero. This leads to two sorts of difficulties. First, if the social discount rate is significantly lower than the market rate of interest, then sustainable development will be economically inefficient and won't be reached in a decentralized market economy. Second, if a single, but a very low discount rate is used for discounting future commodities, then economic agents put a high value on the long distant future. Therefore, optimal intertemporal development will exhibit an unrealistically rapid step-up in near term conventional investment.

3. Modeling

To present ideas in a simple transparent way, a small scale model of global climate change is developed. Time is taken as discrete and each time period is one decade in length. The world is described as though it were a single region operating as a competitive market economy. Among the various greenhouse gases, carbon dioxide (CO_2) is considered as the most relevant one. Potential global warming is caused by increased atmospheric CO_2-concentration and directly affects production, but not utilities.

3.1 Production and climate change

For convenience, let the conventionally measured gross world output y(t) be a Cobb-Douglas function

$$y(t) = \beta L(t)^{\alpha} K(t)^{1-\alpha} \qquad (3.1)$$

of labor L(t) and capital K(t) inputs, respectively. β is a scaling parameter and α is the value share of labor.

To sidestep a detailed energy sub-model, carbon dioxide emissions are viewed as proportional to total output. It is supposed that without greenhouse gas abatement, σ units of CO_2-emissions are emitted if one unit of conventional gross output is produced. Emissions can be reduced, however, by employing abatement activities. Therefore, instantaneous carbon dioxide emissions s(t) are given by

$$s(t) = [1 - a(t)]\sigma y(t), \qquad (3.2)$$

where a(t) denotes the fraction of gross emissions abated in period t. Note that $0 \leq a(t) \leq 1$ is defined as ratio of abated emissions to gross emissions.

As in Nordhaus and Yang (1996) abatement costs as a fraction of gross world output are taken to be quadratic in abatement:

$$m(t) = \tau[a(t)]^2. \qquad (3.3)$$

The scaling factor τ is chosen such that complete elimination of CO_2-emissions consumes 20 % of the world' gross output. This is quite a pessimistic estimate. Suppose for example, carbon-free backstop technologies were available at marginal costs of US$ 200 (see Manne 1996), then the scaling factor τ would be only 5 %. Our estimate is equivalent to average costs of 800$ US and marginal costs of 1400 $ US for carbon free energy.

At any point of time t, the concentration of atmospheric carbon-dioxide Q(t) depends on the former one Q(t-1) and last-period emissions s(t-1). As in Nordhaus (1991), this stock-flow relationship is defined by a first-order difference equation:

$$Q(t) = \Psi Q(t-1) + \Theta s(t-1). \qquad (3.4)$$

This describes two important facts. First, only a fraction Θ of the past period's emissions contributes to the actual atmospheric CO_2-concentration Q(t). The rest is immediately taken up by the oceans. Second, due to long-run mixing processes of surface and deep ocean water, atmospheric CO_2-stocks decline at the rate Ψ over time.

As in Manne (1996), the impact of atmospheric CO_2-concentration on gross production is specified by a quadratic concentration-damage function:

$$\Phi(t) = 1 - ((Q(t)-Q^*)/\Omega)^2. \qquad (3.5)$$

$\Phi(t)$ is the so-called environmental loss factor, i.e., the fraction $1-\Phi(t)$ of the gross world production is lost because of global warming. Ω marks the critical value of the CO_2-stock. At this atmospheric CO_2-concentration, production is reduced to zero. Pre-industrial atmospheric CO_2-concentration (280 ppm) is denoted by Q^*. At this concentration level no climate induced damages are observed.

Multiplication of $\Phi(t)$ and conventional gross output $y(t)$ yields the available 'green output' $\Phi(t)y(t)$

$$\Phi(t)y(t)\,[1\text{-}m(t)] \geq c(t) + b(t) \qquad\qquad (3.6)$$

which can be consumed, $c(t)$, invested, $b(t)$, or used for CO_2 abatement $m(t)$.

3.2 Consumption, climate policy and discounting

We distinguish between two polar opposite cases: one in which individuals do not act altruistically, but where an international institution (e.g. a global environmental bank) sells and buys carbon rights on competitive markets; and one, in which Pareto-efficient climate policies are determined through cost-benefit considerations by an altruistic agent.

In the first case, an overlapping generations model without bequest motive is the best suited approach. However, if generations behave altruistically on behalf of future generations, then in the spirit of Solow (1986), it is reasonable to assume that they have committed themselves to the decisions of an benevolent eco-dictator.

3.2.1 Climate policy in a decentralized overlapping generations economy

As in Stephan, Müller-Fürstenberger and Previdoli (1996) consumers are represented by a sequence of overlapping age cohorts. Each generation passes three phases of life. During the first phase, there are no expenses and no income. During the working phase, capital is accumulated. During retirement there are consumption expenditures but no labor income.

At each date, population is represented by a young, an adult and an old consumer. We have made computational experiments with different, more realistic age-structures. It then turned out that results do not change qualitatively. The macroeconomic gross investment rate is invariant with respect to the age structure. It depends primarily upon the capital-output ratio, the net growth rate of the economy and the rate of depreciation of the capital stock. Obviously, these values are independent of a specific age structure of generations.

Age cohorts are indexed by the date at which they enter active economic live. Since individuals cannot provide for the climate of their offspring acting individually, it is supposed that they have committed themselves to obey the policy-makers' decision to prevent climate damages by controlling annual CO_2-emissions. Note, this policy does not aim for a Pareto-efficient abatement of CO_2-emissions. Instead, the policy goal is to follow an emission trajectory which ensures that global CO_2-concentration are stabilized at 550 ppm by the end of the 21th century. According to that, annual carbon emission rights $\bar{s}(t)$ are issued by an international institution.

Assume, age cohorts maximize life-time utilities and have rational expectations. This means: (1) For each generation t, preferences are represented by utility function $U^t(^1c(t), {}^2c(t+1))$, where $^1c(t)$ denotes consumption during the working period and $^2c(t+1)$ during retirement. (2) Age cohorts correctly forecast market prices for all commodities. Alternatively, there could be a well-defined set of future markets. (For a detailed discussion see Stephan 1995.)

With a three-period OLG-structure, it is reasonable to assume that the young hold the society's endowment of the labor force, l(t), while the old generations own the society's capital endowment. Additionally, let the economically active generations obtain the revenues from selling the carbon emission rights. If p(t), r(t), q(t) and w(t) denote the prices of conventional commodities (output, capital, carbon rights and labor), respectively, then the budget constraint of generation t in its working period is given by:

$$p(t)(^1c(t) + b(t)) \leq w(t)l(t) + q(t)\,\overline{s}(t). \qquad (3.7)$$

The consumer's decision to buy property rights b(t) in future capital is viewed as a savings/investment decision. Together with the depreciation rate λ, the young generation's saving decisions determine the capital stock K(t+1) of the following period:

$$b(t) + (1-\lambda)K(t) = K(t+1), \qquad (3.8)$$

where K(t) is the society's total capital stock at date t.

In the last period of life-time, the old generation has no labor endowment, but receives returns on its capital assets. Therefore, in the period t+1 of retirement the budget constraint

$$p(t+1)^2c(t+1) \leq r(t+1)K(t+1), \qquad (3.9)$$

of generation t $(t = 0,1,2,...)^2$ is now determined only by its capital endowment K(t+1).

3.2.2 Intergenerational altruism

To pin down the concept of intergenerational altruism, we assume that consumption, saving and abatement decisions are delegated to a benevolent policy-maker who maximizes an intergenerational welfare function

$$W = \Sigma_t(1+\delta)^{-t}[U^t(^1c(t), {}^2c(t+1))] \qquad (3.10)$$

subject to the technological, economical and physical constraints (see (3.1) - (3.6) and (3.8)). Total welfare W is defined as the discounted sum of the individual age-cohort's utilities, $U^t(^1c(t), ^2c(t+1))$, as defined in the last section. The discount rate δ reflects the intensity of altruism (see Section 2).

4. Computational Experiments

Computations are based on the Negishi-weight approach (see Manne 1996). They are carried out with GAMS/Minos. Results are reported for a period of 110 years, beginning in 1990. In order to avoid end-of-the-horizon effects, the simulationruns cover 250 years.

We contrast two scenarios: 'Constrained egoism' on one hand and 'intergenerational altruism' on the other. For convenience, the constrained egoism scenario is named 'OLG'. It is based on a decentralized OLG economy. Constraints on annual CO_2-emissions are implemented through tradable emission permits such that the atmospheric CO_2-stock is stabilized at levels below double pre-industrial concentration. The prescribed emission path is derived from an intertemporal optimization approach with very low discount rates (see strong altruism scenario below). Revenues from selling these permits are transferred to the working generations. This ensures maximal investment and growth (see Stephan and Müller-Fürstenberger 1996). During their working period, an age cohort spend its income both on investment and consumption. By contrast, the old generations would spend it completely on consumption (see Section 3.2.1).

For considering the effects of intergenerational altruism on consumption and global climate, pareto-efficient allocations are determined through cost-benefit considerations. This is equivalent to assuming that altruistic individuals have delegated the allocation of resources to a benevolent policy maker who maximizes the sum of discounted welfare. The degree of intergenerational altruism is measured by the utility discount rate δ. We discern three cases: strong ($\delta = 0.5$ %), medium ($\delta = 1$ %) and weak ($\delta = 3$ %) altruism. They are nick-named 'SA, 'MA' and 'WA', respectively.

SA and OLG generate the same time profile of carbon dioxide emissions (see Figure 1, where SA and OLG coincide). In both cases, carbon dioxide concentrations are stabilized at 550 ppm by 2100. Therefore OLG can be considered as the decentralization of a strongly altruistic climate policy in a market economy where individuals follow selfish egoism. WA allows for a significantly higher carbon stock, while MA stabilizes concentration 10 % above SA. Note that without any policy intervention (business as usual) the stock of atmospheric carbon concentration would exceed 1100 ppm at the end of the 22nd century. This is close to the critical value Ω where production is reduced to zero (see equation 3.5).[3]

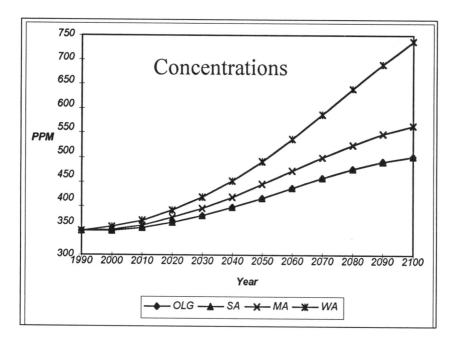

Figure 1: Atmospheric carbon dioxide concentration

For better comparability, the population growth effect is eliminated in Figures 2 and 3. Figure 2 displays the development of per capita green output. Not surprisingly, SA yields the highest green output. OLG and WA coincide for a long period.

Figure 3 shows that OLG and WA are the best policies with respect to per capita consumption. In both cases, per capita consumption is higher than under business as usual conditions, except for the beginning of the time horizon. A closer look at OLG and WA reveals a capital substitution effect: Fig. 2 and Fig. 3 show that green output and consumption do not differ much in the beginnings, but there are significant differences in atmospheric carbon concentration at the end of the time horizon (see Figure 1). Therefore, WA obviously favors man-made capital formation, whereas investment in environmental capital through abatement activities, which pay out in the long-run, are smaller than in the OLG case. However, this difference vanishes beyond 2100.

SA and MA initiate sharp reductions of consumption at the beginning of the time horizon. In both cases high welfare weights are placed on future welfare. Income is withdrawn from consumption in order to build up man-made and natural capital by investment and abatement. Even significant cut backs in current consumption are justified by increased future welfare.

With respect to consumption, WA shows better performance than MA or SA. Discounting utility with 3 % implies a consumption discount rate of 5 %, which is a good approximation of the market discount rate. With 'green' consumption per capita as our criterion, the market discount rate is clearly superior to politically determined low discount rates.

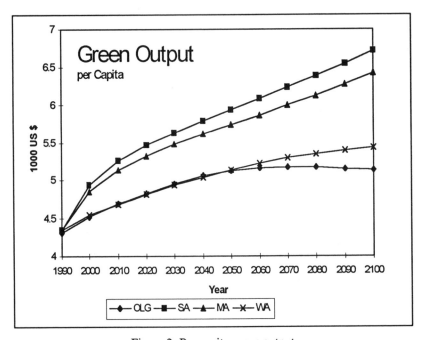

Figure 2: Per capita green output

Comparing the consumption paths in Fig. 3 indicates the difference between short- and long-run effects. In the short run, medium or strong intergenerational altruism hardly pay off. These scenarios reduce current consumption up to 20 % in order to stimulate man-made and natural capital formation. Thirty years from now, the differences become less dramatic. And by the end of the next century, the consumption paths converge.

Figure 4 shows the level of abatement activities. Surprisingly, abatement in OLG exhibits some kind of 'J-effect'. This effect occurs since SA and OLG correspond in emissions but differ in gross output. In contrast to OLG, SA imposes high weights on future generations' welfare. It therefore enforces high near term investment and a sharp rise in gross production compared to the development of green output under OLG assumptions (see Figure 2). Since total carbon dioxide emissions are tied to gross output, OLG requires less abatement then SA to arrive at the same emission levels in any period.

Prices of carbon rights are the shadow prices of carbon dioxide constraints and are related to the marginal costs of abatement (see Figure 5). In the long run carbon prices approach levels at which emissions are avoided almost completely. This can be interpreted as if there were a carbon free energy resource with marginal costs of supply equal to marginal costs of full CO_2-abatement (see Section 3.1).

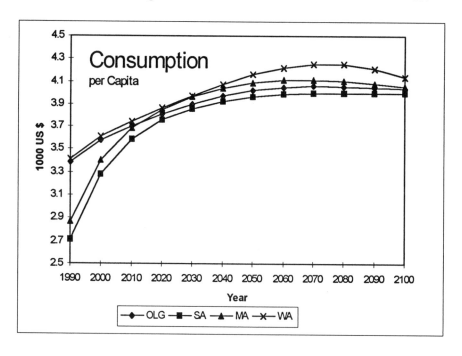

Figure 3: Per capita consumption

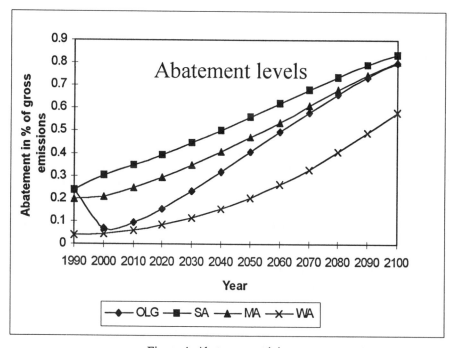

Figure 4: Abatement activity

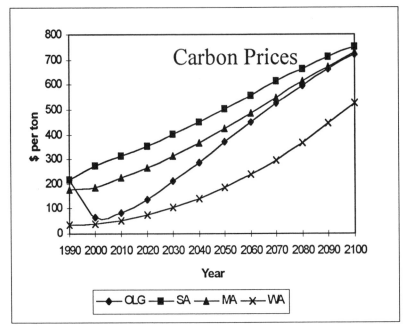

Figure 5: Carbon price

5. Conclusions

This paper was intended to compare prescriptive and descriptive approaches to the issue of sustainability and global climate change. Both approaches can be viewed as polar representations of intergenerational altruism. A prescriptive approach assumes that altruistic agents have delegated the intertemporal allocation of resources to a benevolent policy-maker who acts as a trustee on behalf of both present and future generations.

Using a small scale computable general equilibrium model, we show that strong altruism is required if atmospheric carbon dioxide concentration are to be stabilized at 550 ppm. To meet this ecologically sustainable target, future generations' utilities must be discounted at rates close to zero.

A descriptive approach is best represented within the framework of an OLG economy, where age-cohorts do not care about the welfare of their offspring individually. Nevertheless development can be ecologically sustainable within such a world. This can be achieved through an international institution which enforces an ecological sustainable emission path by means of limiting the sale of carbon permits.

With respect to per capita consumption, a prescriptive approach with strong altruism is clearly inferior to a descriptive one. Strong altruism imposes excessive weights on future generations' welfare. It therefore forces high investments in both natural and conventional capital during the initial periods. These investments cause a sharp cutback of consumption at the beginning of the time-horizon. From a political economy point of view, this is clearly not a feasible option.

6. References

Den Butter, F.A.G. and M.W. Hofkes (1995): "Sustainable Development with Extractive and Non-Extractive Use of the Environment in Production." *Environmental and Resource Economics,* 6:341-358.

Howarth, R.B. and R.B. Norgaard (1992): "Environmental Valuation under Sustainable Development." *American Economic Review,* 80:473-477.

Manne, A.S. (1996): "Equity, Efficiency and Discounting." Mimeo, Department of Operations Research, Stanford University.

Manne, A.S., R. Mendelsohn and R. Richels (1995): "MERGE: A Model for Evaluating Regional and Global Effects of GHG Reduction Policies". *Energy Policy* 23:17-34.

Nordhaus, W.D. (1991): "To Slow or Not to Slow: The Economics of the Greenhouse Effect." *The Economic Journal* 101:920-937.

Norhaus, W.D. and Z. Yang (1996): "A Regional Dynamic General-Equilibrium Model of Alternative Climate-Change Strategies." *American Economic Review,* 86:741-765.

Schelling, T. (1995): "Intergenerational Discounting". *Energy Policy* 23:395-401.

Solow, R.M. (1986): "On the Intergenerational Allocation of Natural Resources", *Scandinavian Journal of Economics* 88:141-149.

Stephan, G. (1995): *Introduction into Capital Theory.* Springer-Verlag Heidelberg u.a.

Stephan, G. and G. Müller-Fürstenberger (1996): "The Double Dividend of Carbon Rights." Mimeo, Department of Applied Micro-Economics, University of Bern.

Stephan, G., G. Müller-Fürstenberger and P. Previdoli (1996): "Overlapping Generations or Infinitely Lived Agents: Intergenerational Altruism and the Economics of Global Warming." to appear in *Environmental and Resource Economics.*

World Commission on Environment and Development (1987): Our Common Future. Oxford University Press, Oxford.

Appendix: Data

Table 1: Benchmark data

Key Economic Variable (1990)	
GDP (trillion US $)	23
Stock of Capital (trillion US $)	63.6
Investments (trillion US $)	4.8
Labor Value Share (%)	72
Labor Efficiency Growth Rate (%) per year	0.5
Population Growth Rate (%) per year	1.5
Depreciation Rate (%) per year	5
CO_2-Emissions (GtC)	6

Table 2: Parameters of the Climate Submodel

Parameter	
Diffusion coefficient per decade Ψ	0.9
Short-run oceanic uptake per decade Θ	0.5
Pre-industrial CO_2 concentration (ppm)	280
Damage due to 2 x pre-industrial CO_2 (% of GDP)	2.5

Endnotes

[1] An earlier version of this paper was presented at an interdisciplinary seminar on sustainability at the University of Berne. Helpful comments provided by the participants are gratefully acknowledged. The authors owe thanks to Alan Manne for his comments, suggestions and stipulating discussion. However the usual disclaimer applies.

[2] Note, t = 0 indicates the old generation. At the beginning of the economy's time horizon, it is in its second period of life.

[3] To keep graphical exposition clear we do not depict business as usual paths in Fig. 1 - 3.

Chapter 9

SEQUENTIAL JOINT MAXIMIZATION

Thomas F. Rutherford
University of Colorado, Boulder

Abstract: *This paper describes a new method by which a competitive market economy may be represented as the optimal solution to a planning problem. The resulting sequential joint maximization (SJM) algorithm solves a sequence of ``partial equilibrium relaxations'' of the underlying general equilibrium model. The partial equilibrium submodels can be solved as nonlinear complementarity problems or as constrained nonlinear optimization problems in either a primal or dual form. This paper introduces the three SJM algorithms, evaluates performance and examines convergence theory. Computational tests demonstrate that SJM is not always as efficient as complementarity methods. A counterexample is presented which demonstrates that local convergence of SJM cannot be guaranteed. Although SJM may have neither the theoretical pedigree of a fixed-point algorithm nor the local convergence rate of a Newton algorithm, the usefulness of the SJM algorithm is demonstrated by the range of real models for which the procedure has been successfully applied. SJM benefits from the availability of robust, large-scale nonlinear programming codes. Consequently, for large scale equilibrium models with inequalities SJM may not be the ``best'' algorithm in theory, but it works extremely well in practice.*

Introduction

There is a close connection between the allocation of a competitive market economy and the optimal solution to a representative agent's planning problem.[1] This paper describes a new method by which this relationship can be exploited in computing equilibrium prices and quantities. This optimization—based algorithm can be applied to find equilibria for models in which consumers have heterogeneous preferences and where production sets may involve point-to-set mappings. Large-scale models may be solved through this technique. Extensions of the algorithm may be applied to models with tax distortions, price rigidities, monopolistic competition and arbitrary preferences. The present paper, however, only considers applications in which producers are perfectly competitive and consumer preferences are homothetic.

John Weyant (ed.), ENERGY AND ENVIRONMENTAL POLICY MODELING. Copyright © 1998. Kluwer Academic Publishers.
ISBN 0-7923-8348-6. All rights reserved.

The algorithm is a Negishi or ``joint maximization" procedure based on a sequence of convex nonlinear programming problems. Typically this sequence will converge to the equilibrium prices and quantities of a competitive market economy. This paper introduces and interprets the algorithm, provides some computational evidence and develops local convergence theory for a simple example.

The idea behind the SJM algorithm is found in the literature concerning conditions under which the demand function describing a set of heterogeneous consumers can be replaced by the demand function for a single agent. In many settings, this is known as the problem of exact aggregation. In neoclassical trade theory this relates to the "existence of community indifference curves". This literature investigates the conditions under which a country's offer curve may be represented as though it arose from a single optimizing agent. In trade theory, at least two such conditions have been identified. The first, due to Samuelson (1956), is that all consumers have identical, homothetic preferences. A second, less widely cited result is due to Eisenberg (1961) and Chipman (1974). This condition places restrictions on endowment vectors in addition to preferences. Eisenberg's result, as summarized by Chipman (1974), is:

> "if each of m individuals has a fixed money income, and if they
> all have homogeneous utility functions (not necessarily identical,
> it must be emphasized), then their aggregate demand function is
> integrable, that is, it may be thought of as resulting from the
> maximization of some fictitious aggregate utility function, subject
> to total expenditure being equal to total income."

In the algorithm described in this paper, Eisenberg's aggregate demand function is embedded in an iterative procedure, which accounts for changes in the relative incomes of agents in the equilibrium system.

The aggregation approach developed by Eisenberg and subsequently interpreted in an economic context by Chipman is presented in a "primal" form. This approach begins with an explicit representation of consumer preferences in which the utility of an individual is specified in terms of quantities consumed of different goods. An analogous result is obtained using preferences characterized in a "dual" form (using the indirect utility function which maps prices and income into utility) which provides an interesting connection between the duality theory of linear programming and the duality of demand function theory. This formulation also renders the computational procedure applicable to models based on econometrically- estimated functions for which primal forms may not exist.

There is an interesting correspondence between the sequential joint maximization (SJM) algorithm and both Mathiesen's (1985) sequential linear complementarity (SLCP) algorithm and Goldsman and Harker's (1990) variational inequality (VI) algorithm for general equilibrium models. SJM can be regarded as a robust form of SLCP; one in which income and price adjustments are effectively decoupled. The SJM algorithm computes a sequence of "partial equilibrium relaxations" of the underlying general equilibrium model. Unlike SLCP (see Mathiesen, 1987), in SJM the choice of numeraire does not affect convergence.

In SJM, subproblems may, in some cases, be guaranteed solvable; however, the outer sequence generated by the algorithm may not converge in all cases. A small example from Manne, Chow and Wilson (1983) demonstrates that when income effects are exceptionally large, a backtracking line search is needed for convergence. A second small example from Scarf (1960) demonstrates that local convergence cannot be guaranteed, even in relatively simple models. In spite of these contrived examples, the experience with large-scale empirical models has been excellent.

Two large scale applications which rely on SJM can be cited.[2] Manne and Richels (1992,1995) use the primal form SJM algorithm to solve ATL, a 13 period, 5 region stochastic decision analysis model of energy-economy interactions, carbon emissions and climate change. The nonlinear programming subproblems for ATL involve 6,800 rows, 8,500 columns and 27,000 nonzeros. The SJM iterations typically converge in 5 major iterations (without a line search) to a satisfactory tolerance. Harrison, Tarr and Rutherford (1992) used the dual SJM algorithm for solving MRT, a 23 commodity, 12 region static international trade model with scale economies and imperfect competition. Monopoly markups and scale economies are handled through Jacobi iterations. Again, convergence involves at most 5 iterations.

Both ATL and MRT are of such size that the complementarity and variational approaches seem infeasible.[3] The fact of the matter is that nonlinear programming solvers like MINOS (Murtaugh and Saunders 1982), CONOPT (Drud 1985, 1994) and CNLP (Kallio and Rosa 1994) are efficient and robust, and SJM makes it possible to use them for equilibrium computations.

The remainder of this paper is organized as follows. Section 2 reviews the formulation of general equilibrium models in a complementarity format, indicating how the equilibrium model reduces to an optimization problem when there is a single household. Section 3 introduces Eisenberg's result in the present context, showing how the multi-agent model with heterogenous, homethetic preferences can be reduced to a single agent model whenever consumers' endowment vectors are strictly proportional. This section also describes the iterative algorithm based on this formulation. Section 4 introduces the dual form of the SJM algorithm. Section 5 relates the SJM algorithm to the complementarity and variational inequality formulations. Section 6 develops necessary conditions for convergence of SJM with Scarf's globally unstable exchange model. Section 7 reports on computational experience with large scale models and concludes.

2. General Equilibrium in a Complementarity Format

Mathiesen (1985) provides a convenient framework in which to represent both partial and general economic equilibrium models. In Mathiesen's formulation, two types of variables characterize an equilibrium: prices, $\pi \in \Re^n$, for n commodities, and activity levels, $y \in \Re^m$, for m production activities. For simplicity, let an activity analysis matrix $A = \{a_{ij}\}$ characterize production possibilities.[4] The element a_{ij} represents the output minus input of commodity i per unit operation of sector j. An equilibrium satisfies two classes of conditions:

1) Perfectly competitive markets assure that no sector earns an excess profit:

$$-A^T \pi \geq 0$$

2) Perfectly flexible prices assure non-positive excess demand for every commodity:

$$A_y - \xi(\pi) \geq 0$$

In the market clearance equation, $\xi(\pi)$ represents the vector of excess consumer demand for market prices π. This vector is the sum of excess demands for each of the consumers: $\xi(\pi) = \Sigma_h (d_h(\pi) - \omega_h)$, where h is the vector of initial endowments for consumer h, and $d_h(\pi)$ is the vector of final demands for consumer h which solves:

$$\max \quad U_h(d)$$

$$\text{s.t. } \pi^T d \leq \pi^T \omega_h$$

Note two features of this equilibrium. First, the equilibrium determines only relative prices because the excess profit equations ($-A^T \pi \geq 0$) and the budget constraints ($\pi^T d \leq \pi^T \omega_h$) are linearly homogeneous in π. For this reason, if (π^*, y^*) constitutes an equilibrium then so does ($\lambda\pi^*, y^*$) for any $\lambda > 0$. The second feature of the equilibrium is complementary slackness. If utility functions are weakly increasing in all inputs and strictly increasing in at least one commodity, then the excess demand vector will obey Walras' law: $\xi(\pi)^T \pi = 0$; and, as a result, equilibrium market excess demands will exhibit complementary slackness with market prices, and equilibrium activity levels will be complementary with unit profits. (Complementary slackness is a feature of the equilibrium, and not an equilibrium condition, per se.)

When there is only one household, the general equilibrium system corresponds to the first-order conditions for the optimization problem:

$$\text{Max } U(d)$$

$$\text{s.t. } \quad d \leq A^T y + \omega$$

Equilibrium prices correspond to the optimal Lagrange multipliers on the system of constraints. The Karush-Kuhn-Tucker conditions impose non-negative profits and complementary slackness on the production activities.

3. Integrable Heterogeneous Demand and a Joint Maximization Algorithm

Let us now consider a model in which all utility functions are homothetic and commodity endowments are strictly proportional. Homotheticity assures that the composition of a utility-maximizing commodity bundle is unaffected by the level of income. As income levels are varied while relative prices remain constant, demand quantities all change in fixed proportions (i.e., the Engel curves are straight lines from the origin). Without loss of generality, we may then assume that utility functions are linearly homogeneous.[5]

Define the expenditure function for household h as:

$$e_h(\pi) = \min \pi^T d$$

$$\text{s.t.} \quad U_h(d) = 1$$

Because $U()$ is linearly homogeneous, the expenditure function provides an index of the "price of a unit of utility". The maximum attainable utility at market prices is then given by the indirect utility function $V_h(\pi) = \pi^T \omega_h / e_h(\pi)$.

Let $\Omega = \Sigma_h \omega_h$ stand for the aggregate (world) endowment vector, and assume that there exists a vector θ for which $\omega_h = \theta_h \Omega$. It can now be shown that the excess demand function representing these preferences and endowments is integrable. Consider the joint maximization problem:

$$\max_{dh} W(U_1(d_1),...., U_H(d_H); \theta) \equiv \Pi \, U_h(d_h)^{\theta h}$$

$$\text{s.t.} \qquad \pi^T \Sigma_h d_h \le \pi^T \Omega$$

where $W()$ is a Cobb-Douglas aggregation function into which are nested the utility functions of the different households. Due to homogeneity, we can reduce this problem to:

$$\max \quad \Pi U_h(d_h)^{\theta h}$$

$$\text{s.t.} \qquad \Sigma_h e_h(\pi) \le \pi^T \Omega$$

where expenditure functions $e_h(\pi)$ are evaluated at prices defined by Lagrange multipliers at the maximum. A global Cobb-Douglas utility function implies demand functions based on fixed expenditure shares for each household. Here, this implies the following relations:[6]

$$\underset{1}{\pi^T d_h} = \underset{2}{e_h(\pi) \, U_h} = \underset{3}{\theta_h \, \pi^T \Omega} = \pi^T (\theta_h \Omega) = \underset{4}{\pi^T \omega_h}$$

In other words, the value of consumer h expenditures equals the value of consumer h endowments. This is true because the Cobb-Douglas value shares in $W(U;\theta)$ correspond exactly to shares of aggregate endowments. It follows that for any price vector π, the demand vector arising from the joint-maximization problem is the same as would result from summing H independent demand vectors. This result corresponds to Theorem 4 in Eisenberg (1961). (See Chipman (1974) for an economic interpretation.)

In practical applications, endowment vectors are typically not proportional. At the same time, it is typical that changes in relative income are small relative to changes in relative prices or quantities. To account for changes in relative shares of aggregate income, the algorithm applies an iterative refinement of the share parameters θ_h.

As an introduction to the algorithm, consider a two-commodity, two- consumer exchange economy in which utility functions have the following (homothetic) form:

$$U_h(x,y) = \alpha_h \, x^{(\sigma h-1)/\sigma h} (1 - \alpha_h) y^{(\sigma h-1)/\sigma h}$$

Each agent has a fixed endowment of the goods x and y, and in equilibrium the sum of demands from agents $h = (A, B) =$ equals the sum of their endowments. Furthermore, at equilibrium prices, the value of each agent's consumption bundle equals the value of endowments.

It is convenient to use an Edgeworth-Bowley box diagram to illustrate both the equilibrium allocations and the joint maximization algorithm. Figure 1 provides a geometric interpretation of the equilibrium. In this diagram, the length of the horizontal axis measures allocations of commodity x and the vertical axis allocations of commodity y. Point e represents initial endowments - consumer A is endowed with (0.9,0.1) units of (x,y) and consumer B is endowed with (0.1,0.9) units; thus there is an economy-wide supply of 1 unit of each good. Point c in figure 1 represents the equilibrium allocation. It lies on the contract curve, the locus of tangency points for agent A and agent B iso-utility curves which runs above the diagonal of the factor box. The contract curve lies above the diagonal because agent A has a stronger preference for y over x as compared with agent B.

The equilibrium point on the contract curve is distinguished as that point from which the line tangent to both agent A and agent B utility surfaces runs directly through the endowment point. Thus, at this point both agents are on their respective budget lines. Point c is an equilibrium allocation for point e as well as for any point along the line from e through c, including point d which lies on the diagonal of the factor box.[7]

Figure 1: Equilibrium in a 2x2 Exchange Model

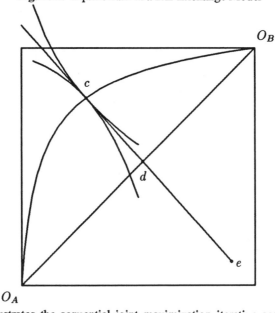

Figure 2 illustrates the sequential joint maximization iterative sequence. The equilibrium values of the Negishi weights are the shares of each agent in the value of global endowments. We begin with a price estimate higher than the equilibrium value, π^0 = 4. These prices define a line (with slope -4) passing through the endowment point e and intersecting the diagonal of the factor box at the point labelled θ_1. This point lies exactly 0.74 of the way from 0_A to 0_B. This ratio represents the value share of agent A endowment as a fraction of the total economy-wide endowment, at the initial price estimate (π^0=4):

$$\theta^1 = \frac{0.9\pi^0 + 0.1}{\pi^0 + 1} + \frac{0.9(4) + 0.1}{(4) + 1} = 0.74$$

Given a value for θ^1, we then solve the joint maximization problem.[8]

$$\max \quad 1.48\log(0.3x_A^{1/2} + 0.7y_A^{1/2}) + 0.78\log(0.8x_B^{1/3} + 0.7y_A^{2/3})$$

$$\text{s.t. } x_A + x_B = 1, \quad y_A + y_B = 1$$

The solution to this optimization problem corresponds to point 1 in Figure 2 (the equilibrium allocation for an economy with endowments θ^1). The ratio of the Lagrange multipliers on the x constraint to the multiplier on the y constraint equals 0.721. This is taken as the price ratio for the start of iteration 2. Continuing in this fashion, we find successively more precise approximations, as indicated by points i_1, i_2 and i_3 in the diagram. Table 1 shows the values of prices, Negishi weights, equilibrium allocations and deviations for 10 iterations of the procedure. Deviations are measured here as the maximum deviation between expenditure and endowment income, as a percentage of endowment income for either agent. In a two agent model, the imbalances for the two agents are always of equal magnitude and opposite sign.

Figure 2: Iterations in the 2x2 model

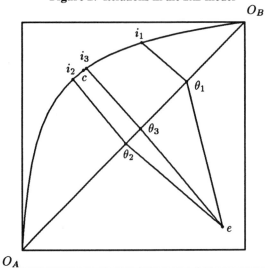

Table 1
Iterative Sequence for the 2x2 Exchange Model

Iter	P	θ_A	X_A	Y_A	Dev
1	4.000	0.740	0.540	0.910	58.300
2	0.850	0.468	0.229	0.751	12.500
3	1.186	0.534	0.290	0.799	3.400
4	1.086	0.517	0.273	0.787	0.900
5	1.111	0.521	0.277	0.790	0.200
6	1.105	0.520	0.276	0.789	0.100
7	1.106	0.520	0.277	0.790	0.000
8	1.106	0.520	0.277	0.790	0.000
9	1.106	0.520	0.277	0.790	0.000
10	1.106	0.520	0.277	0.790	0.000

Key:

     Inter   SJM interation
     P   relative price of X to Y
     θ_A implied income share for consumer A
     X_A, Y_A consumption levels for consumer A
     DEV   percentage deviation in income balance

Algorithm SJM-P - Formal Statement

Initialize: Read an initial estimate of prices, π^0, and set an initial value for the value shares in W:

$$\theta_h^0 = \frac{\pi^0 \omega_h}{\pi^0 \Omega}$$

Set the iteration index, $k \leftarrow 1$, and specify the convergence tolerance (ε) and the minimum step size ($\underline{\lambda}$). Repeat until $\delta(k) < \varepsilon$. In iteration $k = 1,2,\ldots$:

(1.k) Solve:

$$\max \prod_h U_h (d_h)^{\theta_h}$$

(JM-P)

$$\text{s.t.} \quad \Sigma_h d_h \leq A_y$$

(2.k) Read prices from the Lagrange multipliers, $\pi(\theta^k)$.

(3.k) Evaluate the deviation, defined as the largest percentage deviation in any agent's budget constraint:

$$\delta(\theta^k) = \max \left| \frac{\pi(\theta^k)^T \left[d_h (\theta^k) - \omega_h \right]}{\pi(\theta^k)^T \omega_h} \right| \quad x \quad 100$$

(4.k) Perform an Armijo line search (optional)

Initialize $\lambda = 1$, and define:

$$\Theta_h = \frac{\pi(\theta^k)^T \omega_h}{\pi(\theta^k)^T \Omega}$$

If $k = 1$ or $\delta(\theta^k) < \delta(\theta^{k-1})$,

set $\theta^{k+1} = \theta$,
$k \leftarrow k + 1$
return to step (1.k)

else

repeat:

$\lambda \leftarrow \lambda/2$
$\theta = \lambda \theta + (1-\lambda)\theta^k$

solve **JM-P** in order to compute $\delta(\theta)$

    if $\lambda < \underline{\lambda}$ or $\delta(\theta) < \delta(\theta^k)$

    set $\theta^{k+1} = \theta$
return to step (2.k)

endif

continue

endif

The foregoing describes the algorithm as it may be applied for a "generic" model. Some adjustments of the procedure are required to treat models with features such as ad-valorem taxes, nonlinear production, non- homothetic utility functions, etc. (See the Appendix B for the key ideas.)

4. Joint Maximization with Dual Functions

In economic analysis, preferences may be described by a primal-form utility function ($U_h(d)$) or, equivalently, using a dual-form indirect utility function ($V_h(\pi, M_h)$ which expresses welfare as a function of prices and income)[9].

When preferences are homothetic, the unit expenditure function ($e_h(\pi)$) conveys all of the information concerning the underlying preferences. The joint maximization problem (JM-P) can be represented using the expenditure function in place of the utility function, and in many models the computational complexity of the dual-form model is considerably less than for the primal form model.

Let $\overline{M}_h$ be the current estimate of consumer h income. Consider then the following optimization problem:

$$\max \sum_h \overline{M}_h \log(e_h(\pi)) - \pi^T \Omega$$
$$\text{s.t.} \qquad A^T \pi \leq 0$$

The constraints for this problem are the zero profit conditions for the underlying equilibrium problem. The dual variables associated with these constraints are interpreted as the activity levels. The market clearance conditions for the equilibrium are incorporated in the first-order optimality conditions. To see this, differentiate the Lagrangian with respect to π_i. We have:

$$\sum_h \frac{\overline{M}_h}{e_h(\pi)} \frac{\partial e_h(\pi)}{\partial \pi_i} - \sum_j a_{ij} y_j - \sum_h \omega_h \leq 0$$

Notice that from the definition of the indirect utility function:

$$\frac{\overline{M}_h}{e_h(\pi)} = V_h(\pi, \overline{M}_h)$$

According to Sheppard's lemma (Varian 1991), the demand for good I by consumer h equals the utility level times the gradient of the expenditure function, hence:

$$V_h(\pi, \overline{M}_h)\frac{\partial e_h(\pi)}{\partial \pi_i} = d_{hi}(\pi, \overline{M}_h)$$

Substituting into the first order condition, we see that the optimality conditions for (JM-D) correspond to the market clearance conditions which appear explicitly in (JM-P).

The dual problem is low-dimensional. If we are working with a model in which all goods are demanded by all consumers, problem (JM-D) has n variables and m constraints while (JM-P) involves n x p + m variables and n constraints.[10]

As illustration, consider the two agent exchange example from above. In this model, the primal and dual forms are both available in closed form. We have:

$$e_h(p_x, p_y) = (\alpha_h p_x^{1-\sigma_h} + (1-\alpha_h)p_y^{1-\sigma_h})^{\frac{1}{1-\sigma_h}}$$

so that when $\pi^0 = 4$, the (JM-D) problem is:

$$\max \quad 0.74\log\left(\frac{3}{p_x} + \frac{7}{p_y}\right) + 0.52\log\left(\frac{8}{p_x^{1/2}} + \frac{2}{p_y^{1/2}}\right) - p_x - p_y$$

$$\text{s.t.} \quad p_x \geq 0, p_y \geq 0$$

5. The Connection between SJM, SLCP and VI

This section relates the sequential joint maximization algorithm to SLCP. To make this correspondence, we need to make a minor reformulation of Mathiesen's equilibrium conditions, following Rutherford (1987). Consider formulating the model with three classes of variables: commodity prices (π), activity levels (y) and income levels (M). An equilibrium satisfies three sets of conditions:

1) Zero excess profit: $\quad -A^T\pi \geq 0$
2) Market clearance: $\quad A_y - \xi(\pi, M) \geq 0$
3) Income balance: $\quad M = \omega\pi$

Apart from the intermediate income variables, these conditions are identical to Mathiesen's formulation. The vector of excess consumer demand remains the sum of excess demands for each of the consumers: $\xi(\pi,M) = \Sigma_h (d_h(\pi,M_h) - \omega_h)$, but the demand vectors $d_h(\pi,M_h)$ here solve:

$$\text{Max} \quad U_h(d)$$

$$\text{s.t.} \quad \pi^T d \leq M_h$$

SLCP is a Newton method which takes into account inequality constraints and complementary slackness conditions. In each iteration, the system of nonlinear inequalities is approximated by a first-order Taylor series expansion, and the resulting linear system is solved using Lemke's algorithm (Lemke 1965, Ansteicher, Lee and Rutherford 1992).

SJM may be interpreted as a version of SLCP in which prices and activity levels are solved in a simultaneous nonlinear system of inequalities while income levels are adjusted using Gauss-Seidel iterations. The optimization step of the joint maximization algorithm solves the system of zero profit and market clearance conditions, the "fixed income relaxation." During these computations, the income balance equations are (temporarily) ignored. After computing an equilibrium for this nonlinear inequality system, the income variables can be updated using the resulting market prices. Exact replication of the SJM iterative sequence requires that the nonlinear system of market clearance and zero profit conditions be solved completely between income revisions. In the numerical tests reported below, this algorithm is named SJM-C (sequential joint maximization with complementarity subproblem).

A mixture of the SJM and SLCP approaches is related to the variational inequality algorithm proposed by Goldsman and Harker (1990). In the VI algorithm, a fixed-income relaxation is employed for linearized subproblems while incomes levels are revised in every Newton iteration. Subproblems have the form:

$$-A^T \pi \geq 0 \qquad\qquad \perp y \geq 0$$
$$Ay - \nabla \xi\,(\pi,\overline{M})\pi \geq \xi(\pi,\overline{M}) \perp \pi \geq 0$$

This linear complementarity problem corresponds to the first-order conditions for the following quadratic program:

$$\max b^T \pi + \frac{1}{2}\pi^T Q\pi$$

$$\text{s.t.} \quad A^T \pi \leq 0$$

where $b = \xi(\overline{\pi},\overline{M})$ and $Q = \nabla\xi(\pi,\overline{M})$.

The iterative sequence produced by this sequential quadratic programming algorithm will differ from the VI algorithm because income levels are treated as exogenous in each subproblem. By decoupling income and price adjustments, the required number of Newton iterations may increase relative to SLCP. At the same time, this approach may help to avoid defective subproblems which can plague SLCP. (See Mathiesen 1987.)

6. Convergence

Computational experiments demonstrate that a line search is sometimes necessary for convergence. This is the case for a modified version of the Jack--Sprat problem presented in Manne, Chao and Wilson (1980). When Jack--Sprat is initiated from a particular starting point, the SJM procedure with $\lambda = 1$ (i.e. without a line search) after four iterations begins to cycle between two sets of Negishi weights. When the Armijo line search is applied, the outer iterations converge in five steps. (See model Jack-Sprat in Appendix B.)

The second small model presents more serious theoretical problems for the algorithm. Scarf (1960) provides a model which demonstrates that the SJM *differential process* (for $\lambda \to 0$) may be non-convergent. Furthermore, it can be shown that the conditions for convergence of the Negishi process locally reduce to exactly the same conditions for convergence of a simple Tatonnement adjustment process.

Scarf's model involves an equal number of n consumers and goods. Consumer h is endowed with 1 unit of good h and demands only goods h and h+1. Let d_{ih} represent demand for good i by consumer h. Preferences are represented by constant elasticity of substitution utility functions with the following structure:

$$U_h(d) = (\theta^{1/\sigma} d_{hh}^{\frac{\sigma-1}{\sigma}})^{\frac{\sigma}{\sigma-1}}$$

There are two utility function parameters.[11] . θ can be interpreted as the benchmark value share of good i in consumer i demand , and $1 - \theta$ is then the benchmark value share of good $i + 1$. (Consumer n demands goods n and 1.)

The joint maximization algorithm works with consumer income levels, denoted M_h for household h. These may be normalized so that $\Sigma_h M_h = n$. Market clearing commodity prices are determined given the income levels. Let $\pi_i(M)$ denote the price of good i consistent with income levels $M = \begin{pmatrix} M_1 \\ M_2 \\ \vdots \\ M_h \end{pmatrix}$. Provided that $\sigma > 0$, utility functions exhibit non-satiation, so $\Sigma_i \pi_i = \Sigma_h M_h$.

Let $\xi(p;M)$ denote the market excess demand function for fixed income. Given the special structure of preferences and endowments, this function has the form:

$$\xi_i(p;M) = d_{ii} + d_{i,i-1} - 1$$

Given our definition of p(M), we have $x_i(p(M),M) = 0 \forall_i$.

Let $H_h(M)$ denote the difference between the value of agent h allocated income and the market value of the agent h endowment for prices p(M). That is:

$$H_h(M) = p_h(M) - M_h$$

The structure of demand and endowments assure an equilibrium in which all income levels and prices equal unity. When $M_i^* = 1$ then $\pi_i(M^*) = 1 \; \forall_i$, and

$$H_h(M^*) = 0 \forall_h$$

The SJM algorithm, as the step length goes to zero, represents an *income adjustment process:*

$$M_h = p_h(M) - M_h$$

Let the initial estimate M^0 be selected on the n-simplex (i.e., $\Sigma_h \, M_h^0 = n$). The adjustment process then remains on the n-simplex:

$$\frac{d}{dt} \sum_h M_h = \sum_h M_h = \sum_h (p_{h-M_h}) = \sum_i p_i(M) - \sum_h M_h = 0$$

Local convergence concerns properties of the Jacobian matrix evaluated at the equilibrium point, $\nabla H(M^*) = [H_{ij}]$. This Jacobian has entries, which are defined as follows:

$$H_{ij} \equiv \frac{\partial H_i}{\partial M_j} = \begin{cases} \frac{\partial p_i}{\partial M_i} - 1 & i = j \\ \frac{\partial p_i}{\partial M_j} & i \neq j \end{cases}$$

If all prinicpal minors of $\nabla H(M^*)$ are negative, the income adjustment process is locally convergent. If, however, $\dfrac{\partial p_i}{\partial M_h} > 1$, the process is "locally unstable" -- a small increase from the equilibrium income level for consumer h causes consumer h endowment revenue to increase more than proportionally. When an equilibrium is unique and the process is uninterrupted, then local instability implies global instability.

For this model, the Tatonnement price adjustment process is unstable (in the case $n = 3$) when $\dfrac{\theta}{1-\theta} > \dfrac{1}{1-2\sigma}$ (Scarf, 1960). It is shown in the following that the same condition implies instability for the income adjustment process, even though these algorithms produce different search directions away from a neighborhood of the equilibrium.

The function p(M) is defined implicitly by the equation:

$$\xi(p; \; M) = 0$$

In order to evaluate ∇p at M^*, we make a first-order Taylor series expansion:

$$\nabla_p \, \xi(p;M) \; dp + \nabla_M \, \xi(p;M) \; dM = 0,$$

So

$$\nabla p = \left[\frac{\partial p_i}{\partial M_h} \right] = -\nabla_p^{-1} \nabla_M \xi$$

Given the special structure of $\xi_i(p;M)$, we have:

$$\frac{\partial \xi_i}{\partial p_i} = \frac{\partial d_{ii}}{\partial p_i} + \frac{\partial d_{i,i-1}}{\partial p_i}, \quad \frac{\partial \xi_i}{\partial p_{i-1}} = \frac{\partial d_{i,i-1}}{\partial p_{i-1}}, \quad \text{and} \quad \frac{\partial \xi_i}{\partial p_{i+1}} = \frac{\partial d_{ii}}{\partial p_{i+1}}.$$

If we define the "unit-utility" expenditure function for consumer i as:

$$e_i(p) = (\theta p_i^{1-\sigma} + (1-\theta) p_{i+1}^{1-\sigma})^{\frac{1}{1-\sigma}}$$

demand functions are:

$$d_{ii} = \frac{\theta M_i}{e_i^{1-\sigma} p_i^{\sigma}}, \quad \text{and} \quad d_{i+1,i} = \frac{1-\theta) M_i}{e_i^{1-\sigma} p_{i+1}^{\sigma}}$$

Evaluating gradients at $p^* = 1$, it is apparent that the "fixed income Slutsky matrix" has a tri-diagonal structure:

$$\nabla_p \xi = \begin{bmatrix} \delta & \alpha & 0 & \cdots & 0 & \alpha \\ \alpha & \delta & \alpha & 0 & \vdots & 0 \\ 0 & \alpha & \delta & \alpha & & \vdots \\ \vdots & & & & & \\ 0 & 0 & \cdots & \alpha & \delta & \alpha \\ \alpha & 0 & \cdots & 0 & \alpha & \delta \end{bmatrix}$$

where $\delta = -\sigma - (1-\sigma) [\theta^2 + (1-\theta)^2]$ and $\alpha = -(1-\sigma)\theta(1-\theta)$. The inverse matrix, $\nabla_p \xi^{-1}$, is:

$$\nabla_p^{-1} \xi = \begin{bmatrix} \gamma & \beta & \beta & \cdots & \beta \\ \beta & \gamma & \beta & \cdots & \beta \\ \beta & \beta & \gamma & \cdots & \vdots \\ \vdots & & & \ddots & \\ \beta & & \cdots & \beta & \gamma \end{bmatrix}$$

where $\gamma = \dfrac{-(\alpha + \delta)}{(2\alpha + \delta)(\alpha - \delta)}$ and $\beta = \dfrac{\alpha}{(2\alpha + \delta)(\alpha - \delta)}$.

The gradients of excess demands with respect to income at the equilibrium are:

$$\nabla_M \xi = \begin{bmatrix} \theta & 0 & 0 & \cdots & 1-\theta \\ 1-\theta & \theta & 0 & \cdots & 0 \\ 0 & 1-\theta & \theta & \cdots & \vdots \\ \vdots & & \ddots & & \\ 0 & 0 & 0 & 1-\theta & \theta \end{bmatrix}$$

Hence:

$$\frac{\partial p_i(M)}{\partial M_i} = \theta\delta + (1-\theta)\beta$$

Local instability for $n \geq 3$ therefore implies:

$$\frac{\sigma\theta^2 - \theta^2 - \sigma\theta}{3(\sigma\theta^2 - \theta^2 - \sigma\theta) + 3\theta - 1} \geq 1$$

which reduces to the same condition as Scarf demonstrated for the Tatonnement process when $n = 3$.

Given the equivalent local behavior, one might be lead to believe that the income and price-adjustment processes are identical. This is not true. Only local to the equilibrium, where price effects dominate income effects, do these processes follow the same path. This is apparent in Figure 3 where the two vector fields are superimposed. In this diagram, the tatonnement field becomes more divergent than the SJM field as one moves further from the equilibrium point at the center of the simplex. At the equilibrium point the two fields coincide exactly, as demonstrated above.

Figure 3: Tatonnement and SJM Vector Fields Compared

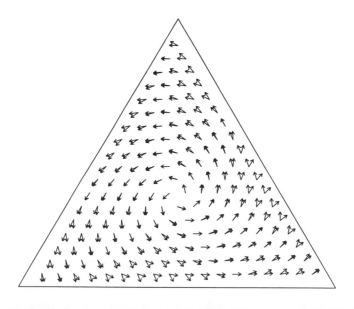

7. Conclusion

This paper has described a new income adjustment procedure which makes it easy to apply the joint maximization algorithm proposed originally by Negishi and Dixon. The paper presented three alternative implementations of the solution procedure, based on primal optimization, dual optimization and complementarity subproblems.

Scarf's model demonstrates that convergence of SJM cannot be guaranteed. At the same time, the SJM approach seems to be the only reliable method for solving large scale models with activity analysis, such as Manne and Richels (1995). Even though SJM has neither the theoretical pedigree of a fixed-point algorithm nor the local convergence properties of a Newton algorithm, the usefulness of the SJM algorithm is demonstrated by the range of real models for which the procedure has been sucessfully applied.

References

Anstreicher, Kurt M., Jon Lee and Thomas F. Rutherford "Crashing a Maximum-Weight Complementary Basis," Mathematical Programming, 54(3), 1992, pp. 281-294.

Carey, Malachy "Integrability and Mathematical Programming Models: A Survey and a Parametric Approach,", Econometrica, 45(8), 1977, pp. 1957-1976.

Chipman, John S. "Homothetic Preferences and Aggregation". Journal of Economic Theory, 8, 1974, pp. 26-38.

Dantzig, G.B., B.C. Eaves, and D. Gale "An Algorithm for the Piecewise Linear Model of Trade and Production with Negative Prices and Bankruptcy", Mathematical Programming, 16(2), 1979, pp. 190-209.

Dirkse, Steven and Michael C. Ferris, "The PATH Solver: A Non-Monotone Stabilization Scheme for Mixed Complementarity Problems", Optimization Methods and Software, 5, 1995, pp. 123-156.

Dixon, Peter J. The Theory of Joint Maximization, North-Holland, 1975.

Drud, Arne Stolbjerg "CONOPT: A GRG Code for Large Sparse Dynamic Nonlinear Optimization Problems", Mathematical Programming 31(2), 1985, pp. 153-191.

Drud, Arne Stolbjerg "CONOPT: A Large Scale GRG Code", ORSA Journal on Computing, 6(2), 1994, pp. 207-216.

Eisenberg, E. "Aggregation of Utility Functions"', Management Science 7(4), 1961, pp. 337-350.

Ginsburgh, Victor .A. and Ludo Van der Heyden, "On Extending the Negishi Appproach to Computing Equilibria: The Case of Government Price Support Poicies", Journal of Economic Theory, 44, 1988, pp. 168-178.

Goldsman, Lynn and Patrick Harker "A Note on Solving General Equilibrium Problems with Variational Inequality Techniques", OR Letters (9), 1990, pp. 335-339.

Harrison, Glenn, Thomas F. Rutherford, and David Tarr "Increased Competition and Completion of the Market in the European Union: Static and Steady-State Effects", Journal of Economic Integration 11(3), forthcoming 1996.

Harrison, Glenn, Thomas F. Rutherford, and David Tarr "Quantifying the Uruguay Round", The Economic Journal, forthcoming 1997.

Kallio, M. and C.H. Rosa "Large-Scale Convex Optimization with Saddle Point Computation", IIASA Working Paper WP-94-107, October, 1994.

Kehoe, Timothy J., David K. Levine and Paul M. Romer "On Characterizing Equilibria of Models with Externalities and Taxes as Solutions to Optimization Problems", Economic Theory 2(1), 1992, pp. 43-68.

Lemke, C.E. "Bimatrix Equilibrium Points and Mathematical Programming", Management Science 11(7), 1965, pp. 681-689.

Manne, Alan .S., Hung-Po Chao, and Robert Wilson, "Computation of Competitive Equilibria by a Sequence of Linear Programs", Econometrica 48(7), 1980, pp. 1595-1615.

Manne, Alan S. and Richard B. Richels, Buying Greenhouse Insurance, MIT Press, 1992.

Manne, Alan S. and Richard Richels "The Greenhouse Debate – Economic Efficiency, Burden Sharing and Hedging Strategies", Energy Journal 16(4), 1995, pp. 1-37.

Manne, Alan and Thomas F. Rutherford (1992) "International Trade, Capital Flows and Sectoral Analysis: Formulation and Solution of Intertemporal Equilibrium Models", New Directions in Computational Economics, W.W. Cooper and A.B. Whinston (editors), 1994, pp. 191-205.

Mathiesen, Lars "Computation of Economic Equilibrium by a Sequence of Linear Complementarity Problems", Mathematical Programming Study 23, North-Holland, 1985, pp. 144-162.

Mathiesen, Lars "An Algorithm Based on a Sequence of Linear Complementarity Problems Applied to a Walrasian Equilibrium Model: An Example", Mathematical Programming 37(1), 1987, pp. 1-18.

Murtaugh, Bruce A. and Michael A. Saunders "A Projected Lagrangian Algorithm and its Implementation for Sparse Nonlinear Constraints", Mathematical Programming Study 16: Algorithms for Constrained Minimization of Smooth Nonlinear Functions, North-Holland, 1982, pp. 84-117.

Negishi, T. "Welfare Economics and the Existence of an Equilibrium for a Competitive Economy", Metroeconomica 12, 1960, pp. 92-97.

Perroni, Carlo "Homothetic Representation of Regular Non-Homothetic Preferences" Economics Letters 40(1), 1992, pp. 19-22.

Rutherford, Thomas F. "A Modeling System for Applied General Equilibrium Analysis", Yale Cowles Foundation Discussion Paper 836, 1987.

Samuelson, Paul A. "Social Indifference Curves", Quarterly Journal of Economics 70(1), 1956.

Scarf, Herbert "Some Examples of Global Instability of the Competitive Equilibrium", International Economic Review 1 (3), 1960, pp. 157-172.

Scarf, Herbert (with Terje Hansen) The Computation of Economic Equilibria, Yale University Press, 1973.

Varian, Hal Microeconomic Analysis, Norton, 1991.

Appendix A

Extensions of the SJM Algorithm

A.1 Nonlinear Production

As in the body of the paper, let $e_h(\pi)$ represent the unit expenditure function for consumer h. Let $\Phi_j(\pi)$ be the analogous "unit revenue function" for sector j, defined as:

$$\Phi_j(\pi) \equiv \max \pi^T x \qquad \text{s.t. } x \in T^j$$

where T_j is the feasible set for the constant returns to scale technology for sector j, operated at unit intensity.

Shepard's lemma characterizes provides optimal producer netput (output minus input per unit activity):

$$x^j(\pi) = \nabla \Phi_j(\pi)$$

A competitive equilibrium is supported by a solution to:

$$\max \sum_h \overline{M}_h \log(e_h(\pi)) - \pi^T \omega_h$$

$$\text{s.t. } \Phi_j(\pi) \leq 0 \ \forall j$$

which has first-order conditions:

$$\sum_j \nabla \phi_j(\pi) y_j + \sum_h \omega_h \geq \sum_h \nabla e_h(\pi) \frac{\overline{M}_h}{e_h(\pi)}$$

A.2 Ad-Valorem Taxes

In the general equilibrium structure, price distortions can, without loss of generality, be applied only to producer inputs.

Suppose that x_j is chosen to solve:

$$\max \ \hat{\pi}^T x$$
$$\text{s.t.} \ x \in T^j$$

in which $\hat{\pi}$ is a vector of tax-distorted prices (users costs), for example: $\hat{\pi}_{ij} = \pi_i(1+t_{ij})$. When tax distortions are present, the tax revenue returned per unit operation of sector j is given by $(\pi - \hat{\pi})^T x$.

In order to accommodate price distortions in the dual joint maximization procedure, the tax distortions are introduced into the constraints, and the tax revenue effects are treated symmetrically with factor endowments - using lagged values for production activities. The generic dual-form optimization problem is:

$$\max \ \sum_h [\overline{M}_h \log(e_h(\pi)) - \pi^T \omega_h \sum_j \tilde{y}_j (\pi - \hat{\pi})^T \tilde{x}^j]$$

$$\text{s.t.} \qquad \Phi_j(\hat{\pi}) \le 0$$

Within the outer loop of the SJM algorithm, income levels are updated by:

$$M_h = \pi^T \omega_h + \sum_j \theta_{jh} \tilde{y}_j (\pi - \hat{\pi})^T \tilde{x}^j$$

in which θ_{jh} is the share of sector j tax revenue which accrues to household h.

Appendix B

GAMS Code

B.1 Two-by-Two Exchange

TITLE: A pure exchange model solved with joint maximization

```
SET    H      HOUSEHOLDS / A, B/

ALIAS (H,HH);

PARAMETER
    EX(H)     ENDOWMENTS OF X /A 0.9, B 0.1/
    EY(H)     ENDOWMENTS OF Y /A 0.1, B 0.9/
    ALPHA(H)  PREFERENCE FOR X /A 0.3, B 0.8/
    SIGMA(H)  ELASTICITY OF SUBSTITUTION /A 2.0, B 1.5/
    RHO(H)    ELASTICITY PARAMETER (PRIMAL FORM)
    THETA(H)  NEGISHI WEIGHTS;

RHO(H) = (SIGMA(H)-1)/SIGMA(H) ;

VARIABLES  U(H)       UTILITY LEVELS
        X(H)       FINAL DEMAND FOR X
        Y(H)       FINAL DEMAND FOR Y
        NEGISHI    NEGISHI ITERATION MAXIMAND

EQUATIONS  OBJDEF       DEFINES THE SJM MAXIMAND
        UDEF(H)      DEFINES UTILITY INDICES
        EQUILX       MARKET CLEARANCE FOR GOOD X
        EQUILY       MARKET CLEARANCE FOR GOOD Y;

OBJDEF..

NEGISHI =E= SUM(H, THETA(H) * LOG(U(H)));

UDEF(H)..
```

U(H) =E= (ALPHA(H)**(1/SIGMA(H)) * X(H)**RHO(H) +
(1-ALPHA(H))**(1/SIGMA(H)) * Y(H)**RHO(H))**(1/RHO(H));

EQUILX..

1 =E= SUM(H, X(H));

EQUILY..

1 =E= SUM(H, Y(H));

MODEL SJM /OBJDEF, UDEF, EQUILX, EQUILY /;

THETA(H) = 1;

U.LO(H) = 0.01;
X.LO(H) = 0.01;
Y.LO(H) = 0.01;

PARAMETER ITLOG   ITERATION LOG;

SCALAR   PXBAR   RELATIVE PRICE OF X IN TERMS OF Y /4/;

SET ITER   NEGISHI ITERATIONS /1*10/;

LOOP(ITER,

    ITLOG(ITER,"P") = PXBAR;

    THETA(H) = (EX(H) * PXBAR + EY(H))
     / SUM(HH, EX(HH) * PXBAR + EY(HH));

    ITLOG(ITER,"THETA") = THETA("A");

    SOLVE SJM USING NLP MAXIMIZING NEGISHI;

    ITLOG(ITER,"X") = X.L("A");
    ITLOG(ITER,"Y") = Y.L("A");

    PXBAR = EQUILX.M / EQUILY.M;

    ITLOG(ITER,"DEV") = ROUND(100 *
    ABS(PXBAR*X.L("A") + Y.L("A")
        - EX("A")*PXBAR - EY("A"))
    / (EX("A") * PXBAR + EY("A")), 1);
);
DISPLAY ITLOG;

B.2 Manne's Jack-Sprat Example

TITLE: The Jack Sprat Problem

*   Prices highly depend on income distribution

* Alan Manne   12/28/91

* The relaxation parameter is theta.

* The SJM procedure cycles when theta = 0, converges

* when theta=0.5.

* From: A. S. Manne, H-P Chao, and R. Wilson:

* Computation of Competitive Equilibria by a Sequence of
* Linear Programs
* Econometrica, Vol. 48, No. 7, November 1980, p 1595-1615.
*
* with a nasty modification of original a(i,j) table.

set i commodities   / 1*5 /
   j activity levels / 1*5 /
   h households   / jack,wife /

table a(i,j) input-output matrix

	1	2	3	4	5
1	-1	.5		-1	-1
2		-1		-1	
3			-1		-1
4			1		
5				1	

* Note: in the version of this problem appearing in MCW,
* the (1,2) coefficient 0.5 is equal to zero.

table b(i,h) endowments

	jack	wife
1	1.	1.
2		1.5
3	1.5	

table u(i,h) requirements for good i per unit of utility

	jack	wife
4	1	
5		1

parameters nwt(h) Negishi weights - initial guess

```
/jack   .1
 wife   .9    /
```

scalar theta relaxation parameter / 0 /

* Turn off solution output for the Negishi iterations:

```
option limrow = 0;
option limcol = 0;
option sysout = off;
option solprint = on;
```

```
variables y(j)    activity levels
         x(h)    utility levels
         gwf     global welfare function
```

positive variables y,x

```
equations sd(i)  supply-demand balance
          gw     global welfare definition;
```

sd(i)..

sum(j, a(i,j)*y(j)) + sum(h, b(i,h)) =e= sum(h, u(i,h)*x(h));

gw..

sum(h, nwt(h)*log(x(h))) =e= gwf

model jacksp / all /;

  x.lo(h) = .01;

* declare Negishi iteration limit

set iter /it1*it10/;

parameters

```
     p(i)          prices
     n(h)          endowment values
     tn            total endowment value
     rwt(iter,h)   revised Negishi weights;
```

* loop over revised Negishi weights

loop(iter,

```
rwt(iter,h) = nwt(h);
solve jacksp using nlp maximizing gwf;

p(i)    = - sd.m(i);
n(h)    =   sum(i, p(i)*b(i,h));
tn      =   sum(h,n(h));
nwt(h)  =   theta*nwt(h) + (1 - theta)*n(h)/tn

);

display p,n,tn,rwt;
```

B.3 Scarf's globally unstable exchange model

TITLE: Scarf's globally unstable exchange model

```
*       This program computes SJM and Tatonnement search
*       directions on a grid over the 3-simplex.

FILE  KSJM /SJM.DAT/; KSJM.NR = 2; KSJM.NW = 15; KSJM.ND = 6;
FILE  KTAT /TAT.DAT/; KTAT.NR = 2; KTAT.NW = 15; KTAT.ND = 6;

SET    ITER   TATONNEMENT ITERATION COUNT /IT1*IT2000/

SET    H      HOUSEHOLDS AND GOODS /G1*G3/;

ALIAS (I,H), (J,H);

PARAMETER
    WEIGHT(H)     INCOME BY HOUSEHOLD,
    PP(I)         PRICE BY COMMODITY (FOR TATONNEMENT)
    EE(I)         EXPENDITURE BY HOUSEHOLD
    X(I)          EXCESS DEMAND;

SCALAR      THETA  OWN GOOD BUDGET SHARE,
            SIGMA  ELASTICITY PARAMETER /0.4/,
            LAMBDA  SPEED OF ADJUSTMENT  /0.1/;

*    SET THETA SO THAT TATONNEMENT IS MARGINALLY UNSTABLE:

THETA = 0.5 / (1 - SIGMA) + 0.15;

ABORT\$(THETA LT 0) " THETA IS LESS THAN ZERO?";
ABORT\$(THETA GT 1) " THETA IS GREATER THAN 1?";

VARIABLE     E(H)  EXPENDITURE FUNCTION
             P(J)  MARKET PRICE
```

OBJ   OBJECTIVE FUNCTION;

EQUATIONS   EDEF(H)
        OBJDEF;

EDEF(H)..

E(H)**(1-SIGMA) =E= THETA * P(H)**(1-SIGMA) +
        (1-THETA) * P(H++1)**(1-SIGMA);

OBJDEF..

OBJ =E= SUM(H, WEIGHT(H) * LOG(E(H))) - SUM(J, P(J));

MODEL   SJM /ALL/;

E.LO(H) = 0.01;
P.LO(H) = 0.01;

E.L(H) = 1;
P.L(H) = 1;

*   EVALUATE DIRECTIONS ON A GRID ON THE 3-SIMPLEX:

SCALAR NGRID   NUMBER OF GRID POINTS FOR FLAGS /20/,
       M   GRID POINTS IN SUBSIMPLEX
       NPOINT FULL SIMPLEX POINT COUNT
       NP   SUBSIMPLEX POINT COUNT
       INDP   PRICE INDEX
       LENGTH FLAG LENGTH;

*   THIS LENGTH SEEMS ABOUT RIGHT:

       LENGTH = 3 / (2 * NGRID);

*   TOTAL NUMBER OF GRID POINTS:

       NPOINT = (NGRID+1) * (NGRID+2) / 2;

LOOP(ITER\$(ORD(ITER) LE NPOINT),

       INDP = ORD(ITER);

*   THESE STATEMENTS GENERATE A GRID ON THE
*   PRICE SIMPLEX:

       M = ROUND(0.5 * SQRT(1.0+8.0*INDP) - 1.5);
       IF (M GT (0.5 * SQRT(1.0+8.0*INDP) - 1.5), M = M - 1);
       NP = (M+1) * (M+2) / 2;
       IF (NP LT INDP, M = M + 1; NP=(M+1)*(M+2)/2;);

```
        WEIGHT("G3") = NGRID - M;
        WEIGHT("G1") = NP - INDP;
        WEIGHT("G2") = NGRID - WEIGHT("G1") - WEIGHT("G3");

*       SCALE TO THE 3-SIMPLEX:

        WEIGHT(I) = WEIGHT(I) / SUM(J, WEIGHT(J));

*       SKIP POINTS ON THE BOUNDARY:

        IF (SMIN(I, WEIGHT(I)) GT 0,

*       Generate the SJM direction:

          PUT KSJM; LOOP(I, PUT WEIGHT(I));
          SOLVE SJM USING NLP MAXIMIZING OBJ;
          PUT KSJM; LOOP(I, PUT P.L(I));  PUT /;

*       Generate the Tatonnement direction:

          PP(I) = 3 * WEIGHT(I);

          EE(H) = (  THETA  * PP(H)**(1-SIGMA) +
            (1-THETA) * PP(H++1)**(1-SIGMA))**(1/(1-SIGMA));

*       EVALUATE EXCESS DEMANDS:

          X(I) = THETA  * (PP(I)/EE(I))
                  * (EE(I)/PP(I))**SIGMA +
              (1-THETA) * (PP(I--1)/EE(I--1))
                  * (EE(I--1)/PP(I))**SIGMA - 1;

*       SCALE ALL THE ARROWS TO BE THE SAME LENGTH:

          LAMBDA = LENGTH / SQRT(SUM(I, X(I) * X(I)));

          PUT KTAT; LOOP(I, PUT (PP(I)/3););
          PP(I) = PP(I) + LAMBDA * X(I);
          PUT KTAT; LOOP(I, PUT (PP(I)/3));  PUT /;

        );
);
```

B.4 Hansen's Activity Analysis Example

TITLE: Joint Maximization: Three Alternative Formulations

\ontext

Reference: Herbert Scarf with Terje Hansen (1973)
    The Computation of Economic Equilibria,
    Yale University Press.

This program generates the following output on a 90 MHz Pentium:

---- 291 PARAMETER ITLOG   Iteration log

	DEV	CPU	ITER
PRIMAL.ITER1	9.8	1.0	243.0
PRIMAL.ITER2	0.9	0.4	22.0
PRIMAL.ITER3	7.748711E-2	0.3	19.0
PRIMAL.ITER4	6.915353E-3	0.3	15.0
PRIMAL.TOTAL		2.0	299.0
DUAL .ITER1	9.8	0.4	40.0
DUAL .ITER2	0.9	0.3	4.0
DUAL .ITER3	7.746462E-2	0.3	4.0
DUAL .ITER4	6.914645E-3	0.3	3.0
DUAL .TOTAL		1.3	51.0
MCP .ITER1	9.8	0.4	14.0
MCP .ITER2	0.9	0.3	2.0
MCP .ITER3	7.746431E-2	0.3	1.0
MCP .ITER4	6.914911E-3	0.3	1.0
MCP .TOTAL		1.3	18.0

For the most part, this is a standard GAMS program. There is one subtlety involved in representing the functions which may appear somewhat obscure to non-GAMS users. In evaluating the SJM maximand, we want to write:

WELFARE =E= SUM(H, THETA(H) * (1/RHO(H))

* LOG(SUM(C, ALPHA(C,H) * (D(C,H)/DBAR(C,H))**RHO(H))));

but we need to account for some special cases. For one thing, if household H has Cobb-Douglas preferences, RHO(H) goes to -INF and the function form (by L'Hopital's lemma) becomes:

SUM(C, ALPHA(C,H) * LOG(D(C,H)/DBAR(C,H)))

To incorporate both functional forms in a single statement, I have used the GAMS dollar operator, writing:

WELFARE =E= SUM(H, THETA(H) *

(((1/RHO(H)) *
  LOG(SUM(C, ALPHA(C,H) * (D(C,H)/DBAR(C,H))**RHO(H)))
)\$(ESUB(H) NE 1) +

```
    (
      SUM(C, ALPHA(C,H) * LOG(D(C,H)/DBAR(C,H)))
    )\$(ESUB(H) EQ 1)
  ) );
```

This simply says, use the CES form if ESUB(H) is not unity, and use Cobb-Douglas if ESUB(H) is unity.

One additional adjustment of the equation is required because some of the share parameters, ALPHA(C,H), may be zero. To avoid evaluating 0/0, I have put another dollar exception operator on the loops over C.

It is an unfortunate fact of life that computers require a precise statement of things, and they get cranky if you ask them to evaluate functions involving 0 times infinity.

\\$offtext

SETS   C   COMMODITIES

  /   AGRIC, FOOD, TEXTILES, HSERV, ENTERT, HOUSEOP, CAPEOP,
      STEEL, COAL, LUMBER, HOUSBOP, CAPBOP, LABOR, EXCHANGE/

      N(C)   NUMERAIRE /LABOR/

      H   CONSUMERS
  /   AGENT1, AGENT2, AGENT3, AGENT4 /

      S   SECTORS
  /   DOM1, DOM2, DOM3, DOM4, DOM5, DOM6, DOM7,
      DOM8, DOM9, DOM10, DOM11, DOM12,
      IMP1, IMP2, IMP3, IMP4, IMP5, IMP6, IMP7,
      EXP1, EXP2, EXP3, EXP4, EXP5, EXP6, EXP7 /

ALIAS (C,CC);

TABLE E(C,H) Commodity endowments

	AGENT1	AGENT2	AGENT3	AGENT4
HOUSBOP	2	0.4		0.8
CAPBOP	3	2		7.5
LABOR	0.6	0.8	1	0.6

TABLE DBAR(C,H) Reference demands

	AGENT1	AGENT2	AGENT3	AGENT4
AGRIC	0.1	0.2	0.3	0.1
FOOD	0.2	0.2	0.2	0.2
TEXTILES	0.1	0.1	0.3	0.1

HSERV	0.1	0.1	0.1	0.1
ENTERT	0.1	0.1	0.1	0.1
HOUSEOP	0.3	0.1		0.1
CAPEOP	0.1	0.2		0.3;

PARAMETER ESUB(H) Elasticities in demand

| / AGENT1 | 1, | AGENT2 | 1, |
| AGENT3 | 1, | AGENT4 | 1 /; |

TABLE IODATA(*,C,S) Activity analysis matrix

	DOM1	DOM2	DOM3	DOM4	DOM5
OUTPUT.AGRIC	5.00				
OUTPUT.FOOD		5.00			
OUTPUT.TEXTILES			2.00		
OUTPUT.HSERV				2.00	
OUTPUT.ENTERT					4.00
OUTPUT.HOUSEOP				0.32	
OUTPUT.CAPEOP	0.40	1.30	1.20		
INPUT .AGRIC		3.50	0.10		0.70
INPUT .FOOD	0.90		0.10		0.80
INPUT .TEXTILES	0.20	0.50		0.10	0.10
INPUT .HSERV	1.00	2.00	2.00		2.00
INPUT .STEEL	0.20	0.40	0.20	0.10	
INPUT .COAL	1.00	0.10	0.10	1.00	
INPUT .LUMBER	0.50	0.40	0.30	0.30	
INPUT .HOUSBOP				0.40	
INPUT .CAPBOP	0.50	1.50	1.50	0.10	0.10
INPUT .LABOR	0.40	0.20	0.20	0.02	0.40

+	DOM6	DOM7	DOM8	DOM9	DOM10
OUTPUT.HOUSEOP	0.80				
OUTPUT.CAPEOP	1.10	6.00	1.80	1.20	0.40
OUTPUT.STEEL			2.00		
OUTPUT.COAL				2.00	
OUTPUT.LUMBER					1.00
INPUT .TEXTILES	0.80	0.40	0.10	0.10	0.10
INPUT .HSERV	0.40	1.80	1.60	0.80	0.20
INPUT .STEEL	1.00	2.00	0.50	0.20	
INPUT .COAL	0.20	1.00	0.20		
INPUT .LUMBER	3.00	0.20	0.20	0.50	
INPUT .CAPBOP	1.50	2.50	2.50	1.50	0.50
INPUT .LABOR	0.30	0.10	0.10	0.40	0.40

+	DOM11	DOM12	IMP1	IMP2	IMP3
OUTPUT.AGRIC			1.00		
OUTPUT.FOOD				1.00	

OUTPUT.TEXTILES					1.00
OUTPUT.HOUSEOP		0.36			
OUTPUT.CAPEOP	0.90				
INPUT .HSERV			0.40	0.20	0.20
INPUT .HOUSBOP		0.40			
INPUT .CAPBOP	1.00		0.20	0.10	0.10
INPUT .LABOR			0.04	0.02	0.02
INPUT .EXCHANGE			0.50	0.40	0.80

+	IMP4	IMP5	IMP6	IMP7	EXP1
OUTPUT.CAPEOP	1.00				
OUTPUT.STEEL		1.00			
OUTPUT.COAL			1.00		
OUTPUT.LUMBER				1.00	
OUTPUT.EXCHANGE					0.50
INPUT .AGRIC					1.00
INPUT .HSERV	0.40	0.40	0.40	0.40	0.20
INPUT .CAPBOP	0.20	0.20	0.20	0.20	0.20
INPUT .LABOR	0.04	0.04	0.04	0.04	0.04
INPUT .EXCHANGE	1.20	0.60	0.70	0.40	

+	EXP2	EXP3	EXP4	EXP5	EXP6
OUTPUT.EXCHANGE	0.40	0.80	1.20	0.60	0.70
INPUT .FOOD	1.00				
INPUT .TEXTILES		1.00			
INPUT .HSERV	0.20	0.20	0.40	0.40	0.40
INPUT .CAPEOP			1.00		
INPUT .STEEL				1.00	
INPUT .COAL					1.00
INPUT .CAPBOP	0.10	0.10	0.20	0.20	0.20
INPUT .LABOR	0.02	0.02	0.04	0.04	0.04

+	EXP7
OUTPUT.EXCHANGE	0.40
INPUT .HSERV	0.40
INPUT .LUMBER	1.00
INPUT .CAPBOP	0.20
INPUT .LABOR	0.04 ;

* DECLARE SETS, PARAMETERS AND EQUATIONS FOR THE SJM ALGORITHM:

SETS   ITER   SJM ITERATION INDEX /ITER1*ITER10/
    STATS   STATISTICS /DEV, CPU, ITER/

PARAMETER

```
        THETA(H)        Budget shares for Negishi model
        P(C)        Market prices
        P0(C)        Initial prices (random starting point)
        INCOME(H)        Endowment income
        EXPEND(H)        Value of allocated demand
        ITLOG        Iteration log;

SCALAR
        DEV        Current deviation
        CONTOL        Convergence tolerance (\%) / 0.01/,
        LAMDA        Damping factor /1/;

ALIAS (C,CC), (H,HH);

PARAMETER
        ALPHA(C,H)        Demand function share parameter,
        IBAR(H)        Reference income associated with DBAR,
        RHO(H)        Primal form elasticity exponent,
        A(C,S)        Activity analysis matrix;

ALPHA(C,H) = DBAR(C,H) / SUM(CC, DBAR(CC,H));
IBAR(H) = SUM(C, DBAR(C,H));
RHO(H) = (ESUB(H) - 1)/ESUB(H);
A(C,S) = IODATA("OUTPUT",C,S) - IODATA("INPUT",C,S);

*=================================================
* PRIMAL MODEL:

VARIABLES     Y(S)        ACTIVITY LEVELS
        D(C,H)        CONSUMPTION LEVELS
        WELFARE        SOCIAL WELFARE INDEX;

POSITIVE VARIABLE Y;

EQUATIONS        MARKET(C)        MARKET CLEARANCE FOR PRIMAL
MODEL
        NEGISHI        MAXIMAND FOR PRIMAL PROBLEM;

MARKET(C)..

SUM(H,D(C,H)) =L= SUM(S,A(C,S) * Y(S)) + SUM(H,E(C,H));

NEGISHI..

WELFARE =E= SUM(H, THETA(H) * (

 ( (1/RHO(H)) *
LOG( SUM(C\$ALPHA(C,H), ALPHA(C,H) * (D(C,H)/DBAR(C,H))**RHO(H)))
 )\$(ESUB(H) NE 1) +
```

```
( SUM(C\$ALPHA(C,H), ALPHA(C,H) * LOG(D(C,H)/DBAR(C,H)))
)\$(ESUB(H) EQ 1)
      ) );
```

MODEL PRIMAL / MARKET, NEGISHI/;

```
*============================================
* DUAL MODEL:
```

VARIABLES   PI(C)   MARKET PRICE
         WELFARE   SOCIAL WELFARE INDEX;

POSITIVE VARIABLE PI;

EQUATIONS   PROFIT(S)   PROFIT CONDITION
         DAFERMOS   MAXIMAND FOR DUAL PROBLEM;

PROFIT(S)..

0 =G= SUM(C, A(C,S) * PI(C));

DAFERMOS..

WELFARE =E= SUM(H, THETA(H) *

```
( ( (1/(1-ESUB(H))) *
   LOG( SUM(C\$ALPHA(C,H), ALPHA(C,H) * PI(C)**(1-ESUB(H))))
 )\$(ESUB(H) NE 1)
```

```
+ ( SUM(C\$ALPHA(C,H), ALPHA(C,H) * LOG(PI(C)) )
 )\$(ESUB(H) EQ 1)
)
```

- SUM(C, PI(C) * E(C,H)));

MODEL DUAL   / PROFIT, DAFERMOS/;

```
*============================================
* COMPLEMENTARITY MODEL:
```

VARIABLES
      PI(C)   MARKET PRICES
      Y(S)   ACTIVITY LEVELS

EQUATIONS
      MCPMKT(C)   MARKET CLEARANCE (FOR THE MCP MODEL);

MCPMKT(C)..

```
SUM(S, Y(S) * A(C,S)) + SUM(H, E(C,H)) =G=
  SUM(H,  ( ALPHA(C,H) * INCOME(H) / PI(C)
  )\$(ESUB(H) EQ 1)

+ ( DBAR(C,H)*INCOME(H)*
  SUM(CC\$ALPHA(CC,H),
      ALPHA(CC,H)*PI(CC)**(1-ESUB(H)))*(1/PI(C))**ESUB(H)
  )\$(ESUB(H) NE 1) );

MODEL MCP   / PROFIT.Y, MCPMKT.PI/;

*=====================================================

* Randomly chosen starting point:

OPTION SEED=1001;
P0(C) = UNIFORM(0,2);

*=====================================================
* Run the primal SJM algorithm:

P(N) = 1;
P(C) = P0(C);

THETA(H) = SUM(C, P(C) * E(C,H));
INCOME(H) = THETA(H);

D.L(C,H) =
  ( ALPHA(C,H) * INCOME(H) / P(C) )\$(ESUB(H) EQ 1) +
  ( DBAR(C,H)*INCOME(H)*
  SUM(CC\$ALPHA(CC,H),ALPHA(CC,H)*P(CC)**(1-ESUB(H)))
    * (1/P(C))**ESUB(H) )\$(ESUB(H) NE 1);

D.LO(C,H)\$ALPHA(C,H) = 1.E-5;
D.FX(C,H)\$(ALPHA(C,H) EQ 0) = 0;

DEV = +INF;

LOOP(ITER\$(DEV GT CONTOL),

    SOLVE PRIMAL USING NLP MAXIMIZING WELFARE;

    P(C) = MARKET.M(C);
    LOOP(N,   P(C)\$P(N) = P(C) / MARKET.M(N) );

    INCOME(H) = SUM(C, P(C) * E(C,H));
    EXPEND(H) = SUM(C, P(C) * D.L(C,H));
    DEV = 100 * SMAX(H\$INCOME(H), ABS(INCOME(H)-EXPEND(H))
                    / INCOME(H));
```

```
ITLOG("PRIMAL",ITER,"DEV") = DEV;
ITLOG("PRIMAL",ITER,"ITER") = PRIMAL.ITERUSD;
ITLOG("PRIMAL",ITER,"CPU") = PRIMAL.RESUSD;

THETA(H) = LAMDA * INCOME(H) + (1 - LAMDA) * THETA(H);
);
ITLOG("PRIMAL","TOTAL","CPU")
      = SUM(ITER, ITLOG("PRIMAL",ITER,"CPU"));
ITLOG("PRIMAL","TOTAL","ITER")
      = SUM(ITER, ITLOG("PRIMAL",ITER,"ITER"));


*=============================================

* Run the dual SJM algorithm from the same starting point --
* this should generate an identical sequence:

P(N) = 1;
P(C) = P0(C);

THETA(H) = SUM(C, P(C) * E(C,H));
PI.L(C) = P(C);
PI.LO(C)\$SMAX(H, ALPHA(C,H)) = 1.E-5;

DEV = +INF;
LOOP(ITER\$(DEV GT CONTOL),

    SOLVE DUAL USING NLP MAXIMIZING WELFARE;

    P(C) = PI.L(C);
    LOOP(N,  P(C)\$P(N) = P(C) / PI.L(N); );

    INCOME(H) = SUM(C, P(C) * E(C,H));
    EXPEND(H) = THETA(H);
    LOOP(N,   EXPEND(H)\$P(N) = THETA(H) / PI.L(N); );
    DEV = 100 * SMAX(H\$INCOME(H), ABS(INCOME(H)-EXPEND(H))
                    / INCOME(H));

    ITLOG("DUAL",ITER,"DEV") = DEV;
    ITLOG("DUAL",ITER,"ITER") = DUAL.ITERUSD;
    ITLOG("DUAL",ITER,"CPU") = DUAL.RESUSD;

    THETA(H) = LAMDA * INCOME(H) + (1 - LAMDA) * THETA(H);
);

ITLOG("DUAL","TOTAL","CPU")
      = SUM(ITER, ITLOG("DUAL",ITER,"CPU"));
ITLOG("DUAL","TOTAL","ITER")
      = SUM(ITER, ITLOG("DUAL",ITER,"ITER"));
```

```
* Run the complementarity-based SJM algorithm from the same
* starting point -- this should generate an identical sequence:

P(N) = 1;
P(C) = P0(C);
THETA(H) = SUM(C, P(C) * E(C,H));

INCOME(H) = THETA(H);
PI.L(C) = P(C);
PI.LO(C)\$SMAX(H, ALPHA(C,H)) = 1.E-5;

DEV = +INF;
LOOP(ITER\$(DEV GT CONTOL),

      SOLVE MCP USING MCP;

      P(C) = PI.L(C);
      LOOP(N,  P(C)\$P(N) = P(C) / PI.L(N); );

      INCOME(H) = SUM(C, P(C) * E(C,H));
      EXPEND(H) = THETA(H);
      LOOP(N,   EXPEND(H) = THETA(H) / PI.L(N); );
      DEV = 100 * SMAX(H\$INCOME(H), ABS(INCOME(H)-EXPEND(H))
                  / INCOME(H));

      ITLOG("MCP",ITER,"DEV") = DEV;
      ITLOG("MCP",ITER,"ITER") = MCP.ITERUSD;
      ITLOG("MCP",ITER,"CPU") = MCP.RESUSD;

      THETA(H) = LAMDA * INCOME(H) + (1 - LAMDA) * THETA(H);
);

ITLOG("MCP","TOTAL","CPU")
       = SUM(ITER, ITLOG("MCP",ITER,"CPU"));
ITLOG("MCP","TOTAL","ITER")
       = SUM(ITER, ITLOG("MCP",ITER,"ITER"));

* Produce a summary report:

OPTION ITLOG:1:2:1;
DISPLAY ITLOG;
```

Endnotes

[1] The use of an optimization problem to characterize equilibrium allocations in a general equilibrium framework is due to Negishi (1960). His original paper was primarily concerned with optimization as a means of proving existence. Dixon (1975) developed the theory and computational effectiveness of "joint maximization algorithms" for multi-country trade models. A survey of early applications of Negishi's theory is provided by Carey (1977). See

also Dantzig, Eaves and Gale (1979). Extensions of joint maximization methods to account for tariffs and labor market imperfections are provided by Ginsburgh and Van der Heyden (1988) and Kehoe, Levine and Romer (1992).

[2] These following claims amount to "proof by testimonial", a technique commonly employed in medical research.

[3] A recently developed complementarity solver, PATH (Ferris and Dirkse 1993) is promising, but it remains to be seen whether it can outperform the optimization-based SJM algorithm, particularly on sparse, large-scale scale models with activity analysis process modules.

[4] The extension to models with smooth (nonlinear) constant returns to scale production technologies is presented in Appendix A. The activity analysis framework is adopted here to simplify notation and also to emphasize the applicability of this procedure for point-to-set mappings, e.g.staircase- shaped supply curves.

[5] An arbitrary function $f(x)$ is homothetic if and only if there exists a non- decreasing function G such that $g(x) = G(f(x))$ is linearly homogeneous in x, i.e. $g(\lambda x) = \lambda g(x)$. These models are based on ordinal rather than cardinal utility, hence $g(x)$ and $f(x)$ may be used interchangeably. It should be possible to accomodate non-homothetic demand using ideas from Perroni (1992), but I leave this extension for future research.

[6] Justification:
1 : Implied by the definition of $e_h()$
2 : Follows from the Cobb-Douglas structure of the aggregation function.
3 : Follows from linear algebra.
4 : From the definition of ω_h (proportional endowments).

[7] The numerical parameters for this example are $\alpha_A=0.3$, $\alpha_B=0.8$, $\sigma_A=2$ and $\sigma_B=1.5$. The equilibrium price ratio (good x in terms of good y) is 1.127. A GAMS program which illustrates the various SJM formulations of the Arrow-Debreu model is presented in Appendix B.

[8] In forming the maximand, the homothetic utility functions have been scaled monotonically to be rendered linearly homogeneous. Hence, we use

$$\overline{U}_A = U_A^2, \text{ so } \theta \log \overline{U}_A = 2\theta \log U_A, \text{ and } \theta \log \overline{U}_B \quad U_A = 3\theta \log U_B$$

[9] See, for example, Varian (1991). Closed form expressions for U() and V() are not always available. Most "flexible functional forms" have a dual form but no explicit primal form. The constant elasticity functions used in the examples below admit both primal and dual forms.

10 When production functions are nonlinear, there are as many additional choice variables in JM-P as there are nonlinear inputs and outputs.

[11] The parameters of this function correspond to Scarf's parameters a and b (Scarf (1960, page 168)) as: $\sigma = \dfrac{1}{1+\alpha}$ and $\theta = \dfrac{b}{1+b}$

Chapter 10

A BRIEF HISTORY OF THE INTERNATIONAL ENERGY WORKSHOP

Leo Schrattenholzer
IIASA, A-2361 Laxenburg, Austria

Section 1: The International Energy Workshop

The International Energy Workshop (IEW) is an informal network of analysts concerned with international energy issues. Its goals are to compare energy projections and to understand the reasons for diverging views of future developments. The main activity of the IEW is iterative polling of energy consumption, production, and trade. The poll includes questions on crude oil prices and GDP in order to identify major determinants of energy demand and supply. The environmental impact of energy use is reflected in the IEW Poll to the extent that it includes items on energy-related carbon emissions. The poll results and other topics of current interest are discussed at annual meetings attended by many authors of energy projections and other individuals involved in the wide field of energy. The original idea to establish this international network was introduced by Alan Manne; Manne organized the first IEW meeting in December 1981 and has served as its senior co-director since then.

Over the years the IEW Poll responses have been collected in a consistent format. They provide a basis for the analysis of the dynamics of energy projections. Such analysis is sobering at times, but it also identifies major strengths of long-term energy and economic projections. The IEW has thus made a modest contribution to quality control in the field.

John Weyant (ed.), ENERGY AND ENVIRONMENTAL POLICY MODELING. Copyright © 1998. Kluwer Academic Publishers. ISBN 0-7923-8348-6. All rights reserved.

Section 2: The First IEW Meeting

The motivation for the first IEW meeting in December 1981 was partly inspired by the Stanford-based Energy Modeling Forum (EMF). One of EMF's functions is to compare energy projections. Modeling teams closely interact to apply their models to the same topic and using a largely standardized set of assumptions. This type of close cooperation was beyond the reach of this newly established network that aimed at analyzing international energy studies. The IEW has therefore taken the practical approach and asks for the results first and later worries about the method that led to them without attempting to provide feedback.

In late 1981, IIASA's Energy Systems Program published a comprehensive volume summarizing its work during the 1970s. An important part of the program's work was the formulation of five scenarios of global energy demand and supply for the time period 1980-2030 (Häfele, 1981). At that time, the Energy Modeling Forum was in its final phases of concluding EMF-6 on *World Oil*, and the International Energy Agency was about to publish *World Energy Outlook*. The energy projections made by these three organizations were the starting point for comparing the projections of 21 groups at the first IEW meeting.

One problem with comparing different projections of the international oil prices in 1981 was that IIASA's 1980 price was a forecast that had been made before the second oil price jump of 1979/1980; therefore, it was too low to be used at face value. As a partial solution to this problem, the first IEW meeting used index numbers, with 1980 as a base, for reporting future oil prices. This solution, however, did not prevent lively discussions about the big differences between the individual projections. Already at this first meeting, the components for later oil price movements were visible. Oil markets had already begun to soften in 1981; therefore the date when the projections of future prices were made in 1981 was an important consideration. The 13 responses for the projected oil price in the year 2000 covered a range that was wider than 100 to 200 percent of the 1980 price (Manne, 1982).

Section 3: The IEW's "Oil Price Era"

Between 1981 and 1986, the global oil price steadily softened. The oil price dropped to less than 50 percent of its peak value in 1980/1981, and the IEW Poll responses reflected this drop. Figure 1 summarizes 15 years of oil price projections. This graph was often used to provide evidence of the inadequacies of oil price projections in general. This conclusion does not seem valid, however, because much information is lost when only medians are reported. As such, the medians reflect little more than conventional and conservative wisdom. The practice of using median responses may have been prompted by Hotelling's "rule" that the price of an exhaustible resource tends to increase in tandem with the real interest rate. Hotelling is not to blame, however, because his hypothesis was formulated for an ideal situation of competitive markets. At any rate, the median is clearly only a single number, and more indicators must be considered to extract the information provided by the whole set of projections.

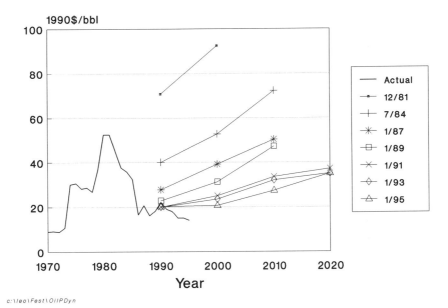

Crude Oil Prices
Actual and Successive IEW Polls

c:\Ieo\Fest\OilPDyn

Figure 1: Crude oil prices: Actual and successive IEW Polls, Manne and Schrattenholzer, 1995.

One tool used to widen the focus of oil price projections was the Random Walk Model (RWM) of International Oil Prices (Manne, 1985). The RWM is the classic stochastic *Random Walk* process consisting of transitional probabilities and a drift parameter. For this presentation, the parameters have been derived from actual oil prices between 1970 and 1995. From this time series of observed values, the average relative increase determines the average increase of the oil price trajectory in the future, and the standard deviation of the relative increases determines the trajectories of the 16% and 84% percentiles, respectively. For the calculation of the standard deviation, the oil price increase of 1973/1974 (by a factor of 2.8) and the oil price decrease of 1985/1986 (by a factor of 2) were eliminated as statistical outliers.

Figure 2 shows the results from the RWM based on these time series. Conceptually, the lower and the upper trajectory of Figure 2 are equivalent to the familiar measurements of uncertainty such as t-statistics, but in the RWM the random element is central and cannot be likened to a "measurement error". In this model uncertainty increases with increasing temporal distance. This feature is in contrast to statistics of IEW Poll responses which sometimes reflected stronger agreement of the projections in the more distant future than in the immediate future. This is also plausible, but it does not lessen the usefulness of the RWM. Oil price forecasting is just another case in which judgement is modified by inconsistent pieces of evidence, each contributing a share to the overall picture.

The Random-Walk Model of the
International Oil Price (A. Manne)

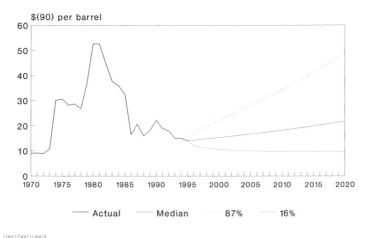

\leo\Fest\rwalk

Figure 2: Oil prices, historical development and projections by the Random Walk Model, Manne, 1985

Originally, mathematical models were introduced into the social sciences to eliminate biases resulting from the response to day-to-day events. Overreacting to new evidence can influence judgmental probabilities over and above a level that would be warranted by a strictly Bayesian analysis. To assess the success of energy models in this regard, we investigate the performance of the different types of models in projecting the international oil price, an area which has been characterized, at least between 1973 and 1986, by a very dynamic development.

To classify modeling tools for projecting the international oil price, we use a table from Nakićenović and Schrattenholzer (1985) which summarizes studies of oil price projections. The studies used for generating the table were conducted in 1982 and 1983. The studies are divided into five groups, and the full range of oil price projections (from the lowest to the highest) in the year 2000 is attributed to each group. Figure 3 is a graphic representation of that table. The number in the parentheses at the base of each column gives the number of reports in that group. Since one report can include many different projections, the table actually summarizes significantly more than 13 (the total of the reports) estimates of the oil price in the year 2000. The following list provides a short characterization of the groups:

(1) *Analytical*. This group includes often econometric-type models that have the oil price as an output.

(2) *Assumption*. Estimates in this group were not explicitly derived in a model-like manner. They may be regarded as educated guesses. This group may be beyond our original subject of inquiry, but there are two reasons for including it: first, it can serve as a useful point for comparison; and second, since this group was included in the original classification, eliminating it might distort the picture.

Oil Price Projections for the
Year 2000, Made in 1983

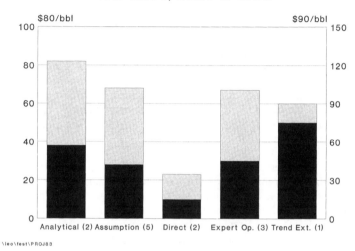

Figure 3: Oil price projections for the year 2000, made in 1982/83, Nakićenović and Schrattenholzer, 1985.

(3) *Direct.* This group includes price-estimating procedures based on estimates of reserves and resources, extraction costs, and demand developments.

(4) *Expert opinion.* There is not much of a difference in either the definition or the outcome between this group and the ``assumption" category.

(5) *Trend extrapolation.* No unique method can be ascribed to trend extrapolation. The classification is just a reflection of the authors' description. A comparison between the ranges and end points of Figure 3 and today's price expectations for the year 2000 suggests that the projections summarized in the *Direct* group are closest to current expectations. Conceivably, this match between projection and outcome could be a coincidence like winning a prize in a lottery. However, a closer look at the 1983 reports in the *Direct* group suggests that reserves and resources figures together with extraction cost estimates provide a more reliable estimate of long-term price developments than econometric methods that rely only on historical time series.

The relative sizes of the ranges also seem noteworthy. As might have been expected, trend extrapolation leads to the narrowest range. The other groups' projections cover the already-discussed range of a factor of two. However, it is important to note that the range of the *Direct* group does not even overlap with any of the other group's ranges.

Section 4: The "CHALLENGE Era"

Between 1991 and 1995, the IEW activities, in general, and the IEW Poll, in particular, were enriched by the results of the CHALLENGE Project. The focus of CHALLENGE was on the analysis of energy-related greenhouse gases emissions and strategies leading to their reduction. National groups generated reference and reduction scenarios, using a small set of standardized assumptions, mainly on oil price development and cutoff points for carbon abatement costs. The global coverage of the CHALLENGE scenarios was more than 90% of the global carbon emissions in 1990. In particular, scenarios for China, the USA, and the former Soviet Union were available.

Based on these results, Manne and Schrattenholzer (1993) produced a "harmonized"reference scenario of the development of global energy-related carbon emissions. The scenario aimed at reproducing the median projections of key variables. To work with a maximum number of inputs, all responses were grouped into one of five world regions. In each group, growth rates of GDP and total primary energy were calculated. Medians of these growth rates were directly used to generate input parameters to the Global 2100 model by Manne and Richels (1992). In addition, other results of the CHALLENGE Poll were used to modify a Reference case that was described in the original Global 2100 model to shape it into the direction of CHALLENGE medians.

On the basis of the Reference case, a Reduction case was calculated in which a carbon tax of US$200 per ton of carbon was introduced. The resulting range between the "do-nothing" scenario emissions and the Carbon Tax case, between 1990 and 2020, is compared with the range between the lowest and the highest carbon emission scenarios of the IPCC (1991). As Figure 4 illustrates, the two ranges overlap, but the Global 2100 CHALLENGE emissions are significantly lower than their IPCC counterparts.

The CHALLENGE Project is expected to end in 1996, but the IEW Poll continues to survey the two items on carbon emissions and reduction. Recent results from OECD carbon emission projections are depicted in Figure 5. The figure shows averages and ranges[1] of this set of IEW Poll responses. As in the RWM, the low curve is the 16% percentile. That is, only one-sixth of the responses falls below it. In particular, a hypothetical stabilization of OECD carbon emission between 1990 and 2000 is clearly outside the IEW Poll range. Such a stabilization is a still widely quoted reduction goal consistent with the U.N. Framework Convention, but the IEW results suggest that it is unlikely to be reached.

Global Carbon Emissions
Global 2100 Scenarios and IPCC Range

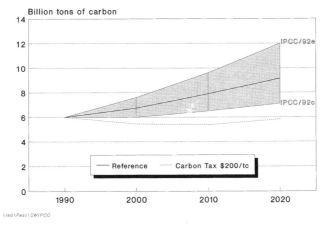

Figure 4: Emission scenarios of IPCC and Global 2100 (CHALLENGE scenarios), billion tons of carbon, Manne and Schrattenholzer, 1993.

OECD Carbon Emissions
IEW Poll'95 Averages and Ranges

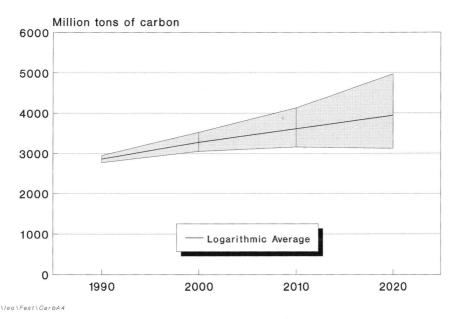

Figure 5: OECD Carbon emissions, averages and ranges, IEW Poll 1995, Manne and Schrattenholzer, 1993.

Section 5: Methodological Considerations

The annual IEW meetings have served as a forum at which methodological reflections on general forecasting issues may be presented. One simple, but powerful, observation was presented by Nordhaus (1985). Basic principles of forecasting efficiency suggest that a projection is just as likely to be on the high side of the actual event as on the low side. A time series of forecasts of the same event that is consistent with this principle would inevitably give a picture of irregular movements. In contrast, a time series of forecasting real U.S. GNP growth for 1982 over 1981 (Figure 6) exhibits a smooth and continuous transition from the initial to the observed value. Nordhaus suggests that people tend to smooth their forecasts too much. That is, they have a conservative tilt that allows them to break the news gradually. Nordhaus appeals to forecasters to incorporate new information quickly.

Forecast of '82 U.S. GNP Growth
Eggers Consensus Forecast

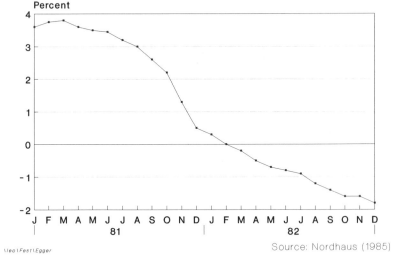

Source: Nordhaus (1985)

Figure 6: Time series of forecasts of US economic growth 1981/82, monthly intervals, quoted by Nordhaus, 1985.

Efficient forecasting is surely of high theoretical value. In practice, however, feedbacks from forecasts are important considerations, in particular from forecasts that are widely publicized. Self-verification and self-falsification are other factors to worry about, even though the effects of these are difficult to measure. Keeping in mind the possibility of such feedbacks increases the recipient's ability to be alert for forecasts that have a clear purpose of generating feedback in these terms.

When we begin to talk about such feedbacks we enter the domain of psychology. To avoid trespassing, we want leave this field to the experts. Articles on heuristics and biases in forecasting are presented in Kahnemann *et al.* (1975). An account of estimates of American oil and gas resources during this century can be found in Wildavsky and Tenenbaum (1981).

References

Energy Modeling Forum, 1982, *World Oil*, EMF Report 6, Summary Report, Stanford University, Stanford, Calif., USA.

Häfele, W. (Program Leader), 1981, *Energy in a Finite World: A Global Systems Analysis*, Vol. 2, Ballinger, Cambridge, Mass., USA.

Intergovernmental Panel on Climate Change (IPCC), 1991, *Climate Change, The IPCC Response Strategies*, Island Press, Washington, DC, USA.

International Energy Agency, 1982, *World Energy Outlook*, OECD/IEA, Paris.

Kahnemann, D., Slovic, P., and Tversky, E. (eds.), 1975, *Judgement under Uncertainty: Heuristics and Biases*, Cambridge University Press, Cambridge.

Manne, A.S., 1982, *International Energy Workshop 1981, Summary Report*, Stanford University Institute for Energy Studies, Stanford, Calif., USA.

Manne, A.S., 1985, *A Random Walk Model of International Oil Prices*, International Energy Project, Stanford University, Stanford, Calif., USA.

Manne, A.S. and Richels, R.G., 1992, *Buying Greenhouse Insurance*, MIT Press, Cambridge, Mass., and London, England.

Manne, A.S. and Schrattenholzer, S., 1993, "Global Scenarios for Carbon Dioxide Emissions," *Energy* 18(12):1207-1222.

Manne, A.S. and Schrattenholzer, S., 1995, *International Energy Workshop, Part I: Overview of Poll Responses; Part II: Frequency Distributions*, International Institute for Applied Systems Analysis, Laxenburg, Austria.

Nakićenović, N. and Schrattenholzer, L. (1985), *The Value of Oil Price Projections*, WP-85-68, International Institute for Applied Systems Analysis, Laxenburg, Austria.

Nordhaus, W., 1985, *Forecasting Efficiency: Concepts and Applications*, Cowles Foundation Discussion Paper No. 774, Cowles Foundation for Research in Economics at Yale University, New Haven, Conn., USA.

Wildavsky, A. and Tenenbaum, E., 1981, *The Politics of Mistrust. Estimating American Oil and Gas Resources*, Sage Publications, Beverly Hills, London.

Endnotes

[1] Ranges are defined here in the same way as in the Random Walk Model described earlier, i.e., as intervals centered at the averages and two sample variances wide.

INDEX